The *Real* Space Cowboys

All rights reserved under article two of the Berne Copyright Convention (1971).
We acknowledge the financial support of the Government of Canada through the Book Publishing Industry Development Program for our publishing activities.

Published by Apogee Books an imprint of Collector's Guide Publishing Inc., Box 62034, Burlington, Ontario, Canada, L7R 4K2, http://www.apogeebooks.com

Printed and bound in Canada

The Real Space Cowboys by Ed Buckbee with Wally Schirra

ISBN 1-894959-21-3 - ISSN 1496-6921

The Real Space Cowboys

Ed Buckbee

with

Wally Schirra

An Apogee Books Publication

Dedicated To Alan B. Shepard, Jr.

"When once you have tasted flight you will always walk the Earth with your eyes turned skyward, for there you have been and there you will always be."
Leonardo da Vinci

Seven brave men volunteered to be first. Together they made 12 epic flights, took a generation of spacecraft into the unknown and pointed the way for generations to follow. They are America's Mercury 7 astronauts. The trail they blazed shines for others, testament to what a nation can accomplish and a measure of man's will. They are part of a brotherhood, joined not by blood, but by mutual respect, national achievement and sincere affection. They are *The Real Space Cowboys.*

This book is dedicated to Alan Bartlett Shepard, Jr., an American hero who gave his all to flying. From pushing the atmospheric envelope as a naval aviator, to brushing the stars during Project Mercury, to an Earth-lit stroll on the lunar surface as an Apollo moonwalker, the unflappable Shepard was driven to succeed. A symbol of America's courage and technology know-how during the Cold War, Shepard was most proud of being selected first from among the elite pool of Mercury candidates. Not one to shy away from questions about his skills, when asked why he was first, he responded, "Because I was the best." That was Shepard. But the "icy commander" was just part of Shepard's persona. There was the fun-loving Shepard; the prankster; the fast talker and fast driver; the man who was frequently discomfited by his celebrity status yet wouldn't hesitate to take advantage of it, especially if it meant a good "gotcha."

His best friend and staunch competitor for four decades, Wally Schirra said, "I voted for Alan B. then (1961) and I would vote for him now, to be the first to climb aboard a U.S. rocket."

Shepard once said, "I died the day Neil walked on the moon." Shepard and others in the brotherhood thought a precedent had been set and that he, not Neil Armstrong, would be the first man to walk on the moon. Unfortunately, a medical condition knocked him from the running. He did become the fifth man to walk on the moon and the only Mercury astronaut to make the trip.

Shepard's desire to extend a hand to the next generation started when he organized the Mercury astronauts to form a foundation with the mission of awarding scholarships to deserving college students. It continued through his support of Space Camp, creating a "little league" of future astronauts and space explorers.

I am greatly indebted to Wally "Skyray" Schirra for recalling events—happy and sad—sharing gotchas, personal experiences that only he could tell; the real stories of the original Mercury astronauts. He makes the book timely, creditable and hopefully, informative and fun.

My sincere thanks to Holly McClain, one of the finest wordsmiths and editors I've ever met, for the hours she spent helping me to organize this story.

Finally, to Gayle, my wife, and our daughters Jana, Jill and Jackie who patiently endured a wannabe astronaut for decades, I offer eternal gratitude. To my grandchildren Zachary, Lane, Matthew, Patrick, Eliza and Knox, you are the future—and my hope.

Mercury astronaut Alan Shepard, author Ed Buckbee and Mercury astronaut Wally Schirra worked together at NASA in the early days of the space program. They were the shakers and movers that started Space Camp, the Astronaut Scholarship Foundation and the Astronaut Hall of Fame. Shepard is a prominent figure in the book as Schirra and Buckbee relate stories about their best friend, the early manned space flight and the Turtle Club founded by Shepard and Schirra.

Acknowledgments

The author would like to acknowledge Sylvia Hundley, Alan Shepard's long time secretary who encouraged me to tell the "real stories" and who departed this life much too early. We miss her. A special thank you to Laura Shepard Churchley who has been most gracious in her kind assistance. I appreciate the stories and material shared by many over the years. I know I have forgotten some of you and please forgive me but I wish to acknowledge the contributions with gratitude of: Pete Cobun, Doris Hunter, Patrick Bailey, Henri Landwirth, Valerie Mayton, Earl Rogers, BobWard, Linda Burroughs, Lee Sentell, Chuck Biggs, Ed Gibson, Claire Johnson, Gene Cernan, Alan Bean, Pete Conrad, Owen Garriott, Marilyn Meyers, Buzz Aldrin, Deborah Barnhart and two outstanding public affairs officers who are no longer with us, Julian Scheer and Dave Harris.

Credits

Thanks to the following organizations and individuals for their support in providing photography, art, graphics, video and DVD's: U.S. Army, Redstone Arsenal, NASA-Marshall Space Flight Center, NASA-Johnson Space Center, NASA-Kennedy Space Center, NASA-MSFC Retirees Association, U. S. Space Camp, U. S. Space & Rocket Center, U.S. Astronaut Hall of Fame, Naval Aviation Museum Foundation, Astronaut Scholarship Foundation, U.S. Air Force, University of Alabama- Huntsville, Coca-Cola, USA, Good Morning America, Regis and Kathy Lee, Mike Douglas Show, Walt Disney Productions, Dinah Shore Show, Late Night With David Letterman, ABC Motion Pictures, 20th Century Fox Television, The Lost Spacecraft, Discovery Channel, Huntsville Educational TV, Keith Ward, WAAF Channel 48, Intergraph Corporation, St. Martin's Press, Tom Fricker, Lisa Fricker, Farley Vaughn, Tony Triolo, Dave Dieter, Jim Taylor, Dudley Campbell, Jim Tuten, Al Whitaker, Sam Boyd and Bob Gathany. Special thanks to the following for their individual efforts: Mike Gentry, Mike Baker, Claus Martel, Daniel Gruenbaum, Linn LeBlanc, Ed Goldstein, Joe Cleemann, Rodney Grubbs, Mike Arrington, Anthony Orton, Dom Amatore, James Bilbrey, Emmett Given, Bob Jaques and Derald Morgan.

Contents

"The Soul of an Explorer Lives In Us All."

… the one significant thing was that we had seven guys all competing for every seat and every flight and we were very strong competitors. But when it got to the point where a guy was assigned to do a flight, everybody folded in behind him and gave 100% support for every flight and every guy.
Deke Slayton

Alan B. Shepard, Jr., and his military buddies were officially presented to me and the rest of the world as America's first astronauts in 1959. The announcement of the Mercury Seven astronauts was made at a press conference at the National Aeronautics and Space Administration's (NASA) temporary headquarters at the Dolly Madison House in Washington, D.C. *"It is my pleasure to introduce to you our astronauts,"* intoned T. Keith Glennan, NASA's administrator, *"and I consider it a very real honor, gentleman: Malcolm Scott Carpenter, Leroy Gordon Cooper, John Herschel Glenn, Jr., Virgil L. Grissom, Walter M. Schirra, Jr., (Wally's name was mispronounced "Scherer") Alan B. Shepard, Jr., and Donald K. Slayton."*

And so began an era of cocky, strutting, swaggering, overconfident military pilots who became overnight, national heroes. *"And we hadn't done anything yet,"* observed Shepard. Each one (except Slayton, who was grounded because of a previously undetected heart condition, but later flew as a crew member of the Apollo Soyuz Test Project) would become a genuine star voyager in Project Mercury. During the five-year life of the project, six human-tended flights and eight automated flights were completed, proving man's adaptability to space. The success of Mercury paved the way for the Gemini and Apollo programs, as well as for all manned spaceflight.

M7's first press conference, Washington, DC in 1959. When asked by a reporter who is ready to go, they all raised their hands--Schirra and Glenn--raised both hands. They are, left: Slayton, Shepard, Schirra, Grissom, Glenn, Cooper and Carpenter.

They were seven, the Mercury Seven (M7). America's new heroes who were going to ride fire-spitting rockets to the high ground, that new frontier called space. They were expected to beat the bad guys and make America number one in space. These seven brave men volunteered to be the first. They would lead a world into space and change it forever. Together they would make the first high risk flights, take a generation of spacecraft into the unknown and point the way for generations to follow. They were of that special breed of man who created the mold and set the standard that became the role model for the U.S. astronaut corps. They were *The Real Space Cowboys*.

All seven were over-achievers with egos to match. With thousands of hours of jet time, the "right stuff" Mercury guys were the best fighter pilots, the best test pilots on the planet—and they didn't mind telling you. Certainly, like all fighter pilots their visual acuity was excellent, but there was something more, something startling about their eyes. They each had deep, intelligent, piercing eyes that took in everything. Always scanning, always searching beyond the obvious, they wore that look of, "Dare me to do it."

Each had a thirst for knowledge, an insatiable curiosity about rockets, space vehicles and anything related to this new field of manned space flight. Indeed, now that they were space flyers, going higher and faster suddenly had infinite possibilities.

The official insignia of the U.S. astronaut corps was designed by Wally Schirra and Gordon Cooper and approved by the M7. The design depicts rocketing to the stars. It is issued to each person who qualifies as an astronaut.

"To hell with the sound barrier—that was Yeager's thing," said Shepard. He and the rest of the M7 wanted to break the bounds of Earth—overcome the gravitational pull of this planet—travel at 17,000 mph, 25 times faster than Chuck Yeager!

The national press corps went wild at the astronauts' coming out party. Scheming, pushing, shoving, cursing and crawling—they fought to get to one of the guys. *"I can't believe it. These people are nuts,"* Shepard asserted. Shepard and company quickly labeled them the "unwashed herd."

The media frenzy, warmed by the public's intense curiosity, brought on a unique contractual relationship between the M7 and *LIFE* magazine. For $70,000, *LIFE* had exclusive news rights to each astronaut and his family; a first for a government-employed test pilot or astronaut.

The M7 continued what was unofficially referred to as the "brotherhood." These were the former military fighter and test pilots who for years competed for flying jobs in a tightly knit, closed culture of male chauvinism. However, the stakes had risen: These jet jockeys were competing for a ride on a rocket into space and to the moon. A fortunate few might actually walk on the moon. This was the ultimate—a chance to ascend to the top of the pyramid.

Thus, the brotherhood of space fliers was formed under the NASA banner. *"Let the games begin. We're going to kick some Commie butt,"* Shepard said. The Russians—referred to euphemistically by Shepard as tractor drivers—were launching satellites, dogs and cosmonauts—male and female—faster then we could record the space firsts.

There was always a bit of a mystery associated with which astronaut was selected for what mission. It would take too many chapters to provide insight into that process, but you can be sure that Deke Slayton, the "godfather of the astronaut corps," not only made the selections that were eventually approved by NASA headquarters, he also named the flights they were on, who during later programs would command the flight, who would walk on the moon and who would stay behind in the mother ship (command module). His fellow Mercury astronaut, Alan Shepard, ably assisted in the selection process. Interestingly, both were on non-flight status because of medical problems when they picked their buddies to go to the moon.

I always wondered why the astronauts did not refer to each other by military rank and call signs since most of them came from the military where they used call signs. Americans really missed something by not hearing Alan "José" Shepard communicating with Jim "Shaky" Lovell or Pete "Tweety" Conrad talking to Al "Beano" Bean or Wally "Sky Ray" Schirra chatting with Tom "Mumbles" Stafford. I think it had something to do with NASA public relations conveying to the American public that this was a civilian program with no ties to the military. Perhaps too, it just wasn't cool for America's heroes to have nicknames. After all, astronauts were star voyagers.

When the M7 astronauts were selected, they had some interesting thoughts about who they were and what was about to happen to them. Thirty years later, those thoughts were captured when Alan Shepard and I asked the guys, to reflect on the past (with the exception of Gus Grissom who lost his life in Apollo 1). The storyline for the nation's first Astronaut Hall of Fame was born. The following are excerpts from those interviews:

What do you recall of those glory days of Mercury when you first started flying in space?

Alan Shepard: *I guess I would have to say the thing that impressed me in the early days and still remains with me is the fact that being an astronaut was always the popular thing to do. I remember there were a lot of people who believed that man shouldn't fly in space. The physiological aspect and psychological concerns about equipment failure were always there. But, I think maybe half of the people thought we were kind of crazy. Maybe more than half thought we were crazy. Most people realized the significance of the exploration and of man flying higher and faster than had ever been done before. Certainly, being an automatic hero was the last thing on my mind in those days.*

John Glenn: *Following up on that, I got a card from the Society of Flat Earth People after my flight with a little letter. All it said was "Dear John, Wise guy!" I think we all recall the specifics of spaceflight and what it felt like to be up there and weightless and things like that... I think about some of the people who were peripheral around the space program and were friends of all of us here, like Leo DeOrsey, the attorney who represented the whole group. Leo tried to get insurance on my first flight and he couldn't. He called up one time and said, "I'm not gonna bet against you on this." He finally found Lloyd's of London that would insure for $100,000 for a six-hour period, but the fee was $16,000. Leo said he wasn't going to bet against me, but he'd write a check for $100,000 and give it to Bill Douglas, our flight surgeon. Leo gave Bill instructions that if I didn't come back, Annie would get that check. That was a real tearjerker I tell you. Then, when I came back from the flight to the Cape, Leo was standing out in front of the Holiday Inn. I stopped the parade to get Leo to ride with us for a little while. Leo said, "I'm sure glad to see you got back!" I think of people like Guenter Wendt, the last hand in the spacecraft before the hatch shut. Many people were part of that whole effort. I think of those people as much as I do the individual, personal experience of being up there and observing things in space.*

Scott Carpenter: *I think about how little we knew during those early days. The first orbital flight by John came back with a real mystery in the form of the fireflies. That one intrigued us all. It turned out they were just little particles of ice, but that simple fact was a great revelation and one of man's many revelations that came out of the early flights. We flew at a time when we really didn't know for sure if the moon wasn't made of green cheese. People forget that. We get complacent; there were a lot of unknowns that are now known because of what we did. There are many unknowns left, but we will learn them.*

Wally Schirra: *Oddly enough, I seem to recall a high-speed training track. Things were happening very, very fast and we had so little time to digest what was going on. It may be what's wrong* with the future, the fact that we were building on something so fast that we really didn't get this organization to build a solid base. As a result, we're finding many problems today; maybe because the enthusiasms and discoveries of the unknowns were mounting so fast we didn't have time to document it all. Now we're sort of looking back and wondering where it all went. I kind of miss what was going on at that time.

Gordon Cooper: *One of the many things I still think back to everyday is sort of a wake up call. How lucky we are to be here! Here we are out on the forefront of things that are going to be, the new way of traveling, and the new way of flying. It's such a privilege to be in this program and be right here on the razor's edge. Everything is unknown and I think it was a real privilege to be in that place and do all that research and step-by-step probe our way into getting man into space.*

Deke Slayton: *Well, since I was the invisible guy in the group and also the only one who didn't fly, I guess my perspectives are a little different. I came out of a military test flight program like everybody else. My biggest shock was to suddenly be popped into this goldfish bowl where everybody in the world was looking at what you were doing. That was a cultural shock to say the least. I'm not sure we've quite adapted to that one, yet! The technical part I never viewed as a big deal and never had any doubt that we were going to be able to do what we said we would. But from the human factor point of view, the one significant thing was that we had seven guys all competing for every seat and every flight and we were very strong competitors. But when it got to the point where a guy was assigned to do a flight, everybody folded in behind him and gave 100% support for every flight and every guy. I think that is significant to the total program's success.*

How did the competition among the M7 manifest?

Slayton: *We were a committee of seven working in one room in those days. Our friends from the press were coming in and trying to interview Alan and John while the rest of us were trying to work some technical problems. It was a little unruly to start but I think it was obvious we all knew we were competing with each other for a job and every guy thought he was best test pilot in the room.*

Shepard: *Still do!*

Slayton: *Everybody thought he was number one and then everyone else was number two.*

Shepard: *I'm not sure I've ever told these guys, but obviously, we all thought we were the best and everybody was totally competitive. Even though we were helping each other from engineering and various operational aspects, there came the time we'd been training almost two years. Late one afternoon, Bob Gilruth said he wanted to see the seven of us in his office. We walked in and closed the door. He said after much deliberation that Shepard would get the first sub-orbital flight, Grissom would get the second and John would be the back-up pilot for both of those flights. Of course, my immediate response was total, pure delight, and then as the guys came forward, it was tough. I knew each one wanted to have that flight and it was a tough moment for me. Talk about mixed emotions.*

Did that somehow separate you from the others for awhile?

Shepard: *About 300 milliseconds!*

Did any of you know each other before you were selected?

Glenn: *Most of us had met at one place or another. I guess I had met Deke, Gordo and Gus at* one time. I had met the others when we were in training at various places. Before that, I had met Scott and Wally, but we didn't know each other real well.

Schirra: *We were ordered to Washington for that first interview, immediately before we went to the Pentagon. We were staying in a tiny Marriott Hotel near the Pentagon. The hotel pond was frozen over and Shepard and I were beginning our careers on icy snow. That was the day we met. The next day, we were given that great privilege to replace the chimpanzee on top of a rocket in a capsule.*

Was there ever a moment when you said to yourself, "Why am I doing this?"

Carpenter: *That thought occurred to me a number of times, generally when I was overly tired. I thought to myself, "What am I doing?" When I got a good night's sleep, I felt differently about it.*

Why would anybody volunteer for something as risky as Project Mercury?

Slayton: *Oh that's very easy. We were in the test flight business when we started and Mercury was a test program and it would be great to do it again. Twenty years after I started this business, I'm still doing it—racing airplanes. So it's just a way of life.*

Carpenter: *Curiosity is the key to that question. It is the prime mover behind every exploration. If there is one thing that can be said to characterize all astronauts, it is curiosity.*

Cooper: *I don't think I felt Mercury was that much riskier than flying a high performance airplane. I think I felt it was going a little farther, faster and higher, and certainly a challenge to be able to get up there.*

Schirra: *I really didn't volunteer for Project Mercury. I was ordered to Washington to listen to a presentation. I had no idea what I was doing there until I saw several test pilots at the same place at the same time, some still very close friends. We were listening to a pair of engineers and a psychologist describing the feeling when you're on top of a rocket in a capsule and going around the world. I was immediately looking for the door and they said, "Not to worry, we'll send a chimpanzee first!" There's no way a test pilot would volunteer for something like that.*

I was encouraged by my peer group, other aviators, successful leaders in the Navy I knew who wanted to go higher, farther and faster and this was the best way to do it. I began to realize that we probably could do that if we could provide input and I maintained this point often. We had to have input because we were not going to go along and fly like a chimpanzee. We forced that input.

Shepard: *I view, as do the other guys, spaceflight and Project Mercury as an extension of what we did primarily as test pilots; notwithstanding any combat experience some had. People don't realize that even back in those days, 30 years ago, we were flying strange looking airplanes, higher and faster than anybody else in the aviation business. Even though the spacecraft didn't look much like an airplane from the outside, from the inside, it looked pretty much like what we flew. At least we tried to make it look like something we understood.*

When you started out in Project Mercury, was the real prize the moon, a pilot's dream of flying higher and faster, or just being first?

Slayton: *I think it was a combination of the last two, being higher and faster and first, since there was only one seat. When we started we weren't going to the moon: We were looking at going into orbit which had not been done.*

Carpenter: *I think each of us had slightly different motivations. Some, I believe, really wanted to be first. Some thought ahead to flying to the moon and some thought of flying one day to Mars. My motivation was not much towards flying higher or faster or being first. I was motivated by curiosity to go to a place where man had never been and bring back understanding.*

Cooper: *Well, from my point of view the goal was to just be a part of the program. I felt very privileged to be in the program. I always felt I'd get whatever flight was meant for me. Initially, of course, we didn't have the goal of going to the moon until well into the program. Then, the moon became a real goal. The goal also was to develop the hardware, equipment and know-how, and find out if we could really get to the moon.*

Schirra: *It* (the moon) *didn't really appeal since I wasn't sure we'd be involved. As you recall, that second group of nine* (astronauts) *was selected and called the Gemini-Apollo group. We were setup as the Mercury group and that was it. The fact that I had a ride on Mercury, Gemini and Apollo truly amazes me. I thought I would be there in the short term and then I thought, "What the heck happened to my Navy career?" It went down the drain when I gave NASA 10 years. I was so immersed in the work of Mercury alone, that I couldn't really get too involved in the lunar mission.*

Shepard: *I guess people really look at the moon differently. If you think in terms of a romanticist, it's a romantic object. Many people seem to feel that maybe we invaded that sort of romantic era. Obviously, the geologist likes to look at the rock formations and how they relate to Earth. Certainly, people who are interested in the money look at what we got back from it. From our standpoint, the moon was just pushing up the frontier. We were going higher, faster and farther and the moon was a nice little target we could shoot for. Here you have something pushing out the frontiers and generating excitement all kind of rolled up into one.*

At that time, just about everything was failing. About the only thing that didn't fail was men. Is that what you were setting out to prove, you couldn't replace men?

Slayton: *No, I don't think so. Although some people thought that was the whole purpose of Mercury—because they didn't have a lot of confidence that man could operate in a micro-gravity environment. Those of us who were flying had no question about man's ability to produce. The real issue was whether the hardware could be made to produce.*

Carpenter: *In a way that's what we all set out to prove. You calculate the risk early on, at the beginning of the training, and you decide whether or not the risk is worth the benefit. If you decide it is, you continue on and forget the risk. You can view the whole operation in flight as an interested bystander. I remember thinking, "I'm interested in the outcome of all of this, but I'm not really involved. I'm watching it as an outsider."*

Cooper: *Well, we set out to prove that a little extra precaution by the Mercury stamp* (of approval) *that went on every major piece of hardware would make a better piece of equipment. By adding inspections and tender loving care at check out and testing, we could build vehicles that were totally reliable. Add man to it, and the back-up system to assist man, and we felt it was going to be a pretty safe system.*

You had rivals on the other side, the Russians. When the Mercury program ended, did you think you were winning or losing the race to the moon with Russia?

Slayton: *I thought we were losing. We were behind them in every area you could imagine. They* *were doing extra-vehicular activities, multi-manned things, and had more hours in space than we did. It occurred to me that we were rather far behind at that point.*

Carpenter: *By the time Mercury ended, we were pulling even. We still had a long way to go, but contrary to the way a lot of people felt in those days, I didn't view the space race as having an end on the moon. There is no end to the space race: It doesn't even end on Mars.*

Cooper: *Well, for starting at least five years behind the Soviets in space, I felt we had really made major strides in Mercury. There was no doubt in my mind that we were going to win the race to the moon. We certainly had no way of proving that because there was always doubt as to exactly where the Russians were in their technology. Almost all of us had confidence that we were going to beat them to the moon.*

Schirra: *We were ahead. Actually, I didn't know this until later, but I was the fifth American to go into space before a fifth Russian had been in space. Just a little bit of trivia. Mercury didn't do what we hoped it would do, which was to give us a long duration flight, much longer than the 34 hours Gordon Cooper did. But when Gemini came along, I was convinced we were on our way to the moon.*

You were military pilots and test pilots. Did you live each day like it was your last? Did you expect the press to treat you like heroes?

Slayton: *That was probably my biggest shock. A bunch of military pilots, and suddenly we were in view of the whole world. Most of us didn't know how to cope with that. I had a tough time. But I learned that most of the press people were trying to do a job and were more likely to help you than hurt you. So a lot of them got to be good friends. That's the way it should be.*

Carpenter: *I suppose we were immediately recognized as American heroes. That turns out to be an occupational hazard for space pioneers. It's okay and has advantages and disadvantages. The disadvantages mainly relate to the loss of anonymity. Anonymity is something you don't really appreciate until you've lost it. But that again is part of the risk/benefit ratio. It was well worth it because spaceflight is a supreme experience.*

Cooper: *Well, according to the public we seemed to be heroes and it was a big surprise to us. It was certainly a surprise to those of us from the Air Force who had never really been exposed to public relations before. It was a changed world. I felt they were a little premature, making such heroes out of us before we'd done anything. I would rather…it had been more low key.*

Schirra: *That was my goal—to be a hot shot test pilot, not just a scarf and goggles type, but one who could use his engineering confidence to work on systems and make the best airplane, ever. Then, be the one responsible for flying that airplane. Seeing all these people excited about these seven guys who hadn't done anything was a little amazing.*

Shepard couldn't believe what was happening to them. The press and the public loved them as brave and courageous men who had volunteered to face the unknown. These astronauts were a fearless breed of hell-raising jet jockeys who hadn't quite grown up. They had a lot of swagger. They existed primarily in an elite, fighter-jock culture and did nothing but breathe, eat, talk, sleep and fly hot jets.

"Shepard had it all, he was the epitome of the American spirit," said Julian Scheer, NASA Public Affairs director during the Apollo era. "He was supremely confident, a confidence that bordered on arrogance. He had a certain rebelliousness as far as authority was concerned. So you had a cocky, self-assured guy, and that was what Americans wanted at the time. There was not a humble bone in Alan Shepard's body, and that's the way he was throughout the entire space program...He was a superb mixture of military demeanor and a rebel on a motorcycle."

Vice-President Lyndon Johnson presents the M7, Carpenter, Cooper, Glenn, Grissom, Schirra, Shepard and Slayton in the approved and accepted order, "CGGSSS" at the Sam Houston Coliseum, Houston, Texas. Shepard added a note to this picture that read, "To Ed Buckbee: did you have a job in '62?"

Man must rise above the Earth—to the top of the atmosphere and beyond—for only thus will he fully understand the world in which he lives. **Socrates, 500 B. C.**

No one in the U.S. had ever trained for spaceflight. The M7 were the first Americans subjected to the rigorous training for high risk flights aboard rockets that could propel man into the weightlessness of outer space.

Little did they know what it would be like to train for these flights. Arduous, physical training was easily anticipated, but the seemingly endless medical tests from the pencil pushing, finger jabbing "medical cult"—a term made popular among the brotherhood—was often beyond their belief. Claiming they had the right to probe and find orifices the astronauts didn't know they had, the physicians and technicians kept the M7 on their toes. When John Glenn was asked what he disliked most about examinations, he answered, *"When they (physicians) went into every opening of our bodies just as far as they could go to probe and collect fluids and other substances."* The medical people, as well as some of the engineers, believed the astronauts to be mere passengers or experiments, and need not be skilled in performance or decision making. Thus, began the trend by some to compare the M7 astronauts to chimpanzees.

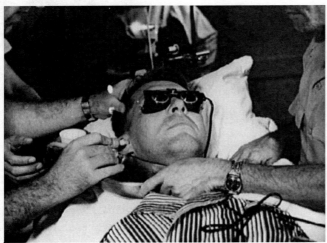
Getting poked by the medical cult, Schirra said, "They found openings in my body I didn't know I had."

As the one-seat Mercury capsule was being built and the two-seat Gemini capsule conceived, each of the M7 was assigned specific areas of responsibility. Each became a subject matter expert on a specific part of the capsule. Collectively, they understood the total system. This enabled the M7 to always take part in any changes that occurred to the capsule.

About this time, I was beginning my first assignment as a U.S. Army officer at Redstone Arsenal, Alabama. I arrived in 1959—right after the boys in the Army and a guy named Wernher von Braun had converted a Redstone ballistic missile to successfully launch the Free World's first satellite, Explorer I.

In 1960 von Braun and his team were transferred from the U.S. Army to the newly created George C. Marshall Space Flight Center under the National Aeronautics and Space Administration (NASA) that Congress had established as the civilian agency to conduct the peaceful exploration of outer space. Von Braun was named the director of the Huntsville-based center. Shortly after that, I resigned my Army commission and went to work as a NASA public relations guy for the von Braun team.

The astronauts who took the first flights into space and made the historic steps on the moon are men I first met in Huntsville, working for von Braun, "father of the moon rocket." As newly selected astronauts, they naturally came to see Huntsville's rocket factory and to meet von Braun, the man who was designing and building a transportation system that would eventually take them to the moon. They climbed test stands, perused instruments units, examined rocket nozzles and bombarded their rocket host with questions.

First meeting of M7 in 1959 with Wernher von Braun and his U.S. Army boss, Maj. Gen. John "Big M" Medaris, at Redstone Arsenal, AL.

Did you build some of your own training equipment? How did astronaut training compare to training to fly airplanes?

Slayton: *In an airplane, we do a high-speed taxi and go up and fly around with the gear down. We build up the flight envelope and then when you get that far along, there's only one good place to go and that's into orbit. We didn't have any way to train for it incrementally except with the simulator.*

Glenn: *We designed it as we went along. We didn't know what kind of training was involved so we tried everything we could think of that might be worthwhile. If anything, we over trained in a lot of different areas and were over-examined. I remember the medical exams and they probed, prodded and ran every medical test they knew how to run on a human body, literally. Psychiatric tests and screenings, we were sort of the guinea pigs obviously. The same thing happened in the training at the Cleveland Lewis lab. I'm sure everybody remembers the old Multiple-Axis Space Test Inertia Facility (MASTIF) which was a three gimbals rig about 20 feet across with a simulated spacecraft in the middle with an array of instruments and a little hand controller. They would turn on different axes, and as you get turned up and roll, you'd learn how to control that and bring it to a stop. Then they'd turn it up in pitch and roll, pitch and yaw, one axis at a time, and then two at a time. Your graduation exercise on that thing was to go at 30 rpm's in roll, pitch and yaw, all at the same time. You talk about a built-in barf machine, that thing is the closest I ever came in the program to losing all my breakfast and other things along with it, I guess. That was one example of where we, perhaps over trained. The idea was to see whether you could control a spacecraft if it went out of control and if the thrusters were turning up at a high rate.*

Shepard: *Neil Armstrong proved that.* (Armstrong saved the Gemini 8 mission when a thruster continued to fire causing the spacecraft to spin out of control.)

Schirra: *One of our friends is Dr. Bill Douglas* (physician for the M7) *who did everything we did but fly. If you really look back on those days when we were being tested, we were healthy specimens being investigated by sick doctors. What I admire about Bill Douglas, and I think we should remember him for, is he tried to make things more comfortable for us. One was to have specially painted four tattooed spots on our bodies… for placing electrocardiogram sensors. Bill Douglas had that done so now we all have our four little dots on us. I think we're unique in that way at least.*

Glenn: *We did work up in Johnsville, Pennsylvania on the human centrifuge. We worked up as high as 16 g's.* (That's 16 times your weight on Earth.) *It's not like flying an airplane, but that's a lot of g's, I tell you. At 16 you're doing everything you can to stay calm. Was that overtraining? I keep looking back on it and it was overtraining, but they were trying to give us an experience of everything* *they could possibly think of. Underwater, we did scuba with frogmen because that was as about as close as we could come to continuously simulating weightlessness, although it's not really the same.*

The M7 trained in the MASTIF-Multi-Axis Space Training Inertial Facility-sometimes referred to as the "barf machine" by John Glenn and other members of the brotherhood.

Carpenter: *Because there were so many unknowns about how the human organism would respond to these new stresses, the spacecraft development itself sort of lagged. We probably were more thoroughly investigated in psychological and physiological ways than any other group of men, ever. The thing that amazed me was that this human organism is so miraculously designed as to be adaptable to these stresses. The human body is a marvelous machine. On the centrifuge, for instance, we had no evolutionary experience with high acceleration. But lo and behold, there is almost an autonomic defense to that high acceleration. You don't need training, but practice helps. You know automatically how to combat it and I think that is an amazing thing. The human body is very well adapted to spaceflight and the rigors with which it has no evolutionary experience. It's a marvelous thing.*

Shepard: *John, do you remember the sign that was over the door of the centrifuge that whirled us around and everything was strained to the limit? There was a big sign over the door saying, "If found please return to Johnsville, Pennsylvania!*

What areas of the spacecraft did each of you become subject matter experts?

Shepard: *Well, we all had our responsibilities. It really was very important, and despite the competitive nature of pilots, brought us together to a great degree. Because of my Navy background and training I was assigned the responsibility of the rescue. As you know, landings were made in the ocean in those days and the Navy was responsible for that. That was my area of responsibility. It also included the escape provisions should the capsule go underwater and need immediate evacuation. Of course, the guys all practiced crawling up to the top of the capsule and going through the helicopter recovery procedures.*

Glenn: *I was responsible for the cockpit layout and the instrumentation. I had done some work in aircraft at Pax (   ) River at the Naval Aviation Test Center. We had a giant argument within our own group over the three-axis hand controller and whether we were going to have rudder pedals. I remember Deke would refer to us as "rudder pedal jockeys." He didn't want to get away from that. It's what you could really control with the three-axis where you controlled roll, pitch and yaw all with a hand controller. We had some really fine ideas on instrument layout, tape line presentations and all. We laid out little mock-ups and then found out with the battery power onboard we couldn't run all those things and still run a mission beyond just a few hours. So, we had to go back to pretty much the* *old steam gauges which didn't use much power, and little electrical gauges that were off-the-shelf type items. Today, when you look at an old Mercury spacecraft you see these things that look even more primitive than we were capable of using and designing at that time. The cockpit layout, what we put where, when you were going to use instruments during what phase of flight, and how you were going to do that, was what I concentrated on.*

The Johnsville centrifuge was another extreme training device used by the M7 that gave the astronauts 16-g's, sixteen times their normal weight.

Carpenter: *My responsibility had to do in part with navigation and communication equipment. I'd flown some Navy airplanes that had equipment similar to our periscope, which was our primary visual navigator. We couldn't navigate in the classic sense and were committed to one course in that machine. We could keep track of our progress through the periscope. That was my area of expertise, but we all crossed over a great deal into all areas of the spacecraft design.*

Schirra: *I approached the training with some experience from my Navy training at Pax River. The pilots in flight wore what we called the full pressurized suit. The Navy had elected to fly a fully pressurized suit in lieu of a partially pressurized suit the Air Force had been using. With that experience, I became the so-called suit expert. Of course, that suit became the ultimate thing that let man walk on the moon. The thing I hated was that dumb group picture of the seven of us wearing those silver suits (in The Right Stuff movie). It was the worst movie that's ever been done. We wore that suit simultaneously for probably a total of one-half hour.*

Shepard: *At that time we only had one (space) suit and we all took turns wearing it!*

Schirra: *The other area I was concerned with other than the suit, of course, was the environmental control system. We were wondering how that would work. The same flight that John and Scott have been talking about, we all saw these floating particles in orbital flight. They were a result of urine wastewater being dumped overboard; that's how we cooled our suit through the environmental system. We boiled water at about 40 degrees Fahrenheit, which is cooled enough to cool us off. As a result, there were molecular ice crystals as they left the spacecraft. During Gemini I saw this thing hanging on the side of Borman's and Lovell's Gemini 7 spacecraft that had been up there for several days. It turned out to be frozen urine that had formed an icicle. We referred to it as the Gemini 7 "pissicle."*

Cooper: *I had a little bit of a propulsion background so the Redstone rocket was primarily my responsibility. As well, I had pad rescue and that included any emergency situation we might have on or around the pad. Interestingly enough, after a very expensive development program, I was pleased I was the one who had to be gotten out in Gemini in an emergency. I finally had a chance to use it.*

Shepard: *You're the one* (speaking to Cooper) *who supported, after a big series of discussions, whether we should have a window or a porthole. Remember that? After long deliberations—because it was going to cost more money and take time—you're the one who really got to use the window as a backup when your gyros failed.*

Schirra: *Talking about the window, I never had a chance to ask Alan on his flight about what he said, "My, what a beautiful view." You didn't have a window or a little porthole. What were you looking at?*

Shepard: *Well, it was actually a taped remark, Wally. You remember how regimented we were on those early flights, particularly on the sub-orbital, which was only a matter of 16 minutes or so? There came a time toward the apogee of the trajectory when... the door came open and I could look down and it was actually aligned vertically. I saw the entire horizon with my hemispherical field of view almost 180 degrees. I looked out and saw the entire horizon and, of course, the Caribbean Islands, the clouds, the ocean and the offshore rigs. It was just an overwhelming view and that's when I said, "Man, what a beautiful view."*

Slayton: *Getting back to your question about experts, we were trying to convert the propulsion systems from different applications into a manned carrier. In my case it was the Atlas, which was an intercontinental ballistic missile. It was a big step from there to make it a manned carrier. We had developed an escape system on the thing and spent a lot of time around the factory convincing workers that we were really dealing in a manned system. It's significant to note that when John flew the first Atlas we had a probability of failure of one in five. On the heels of the Challenger thing, people talked about how unsafe space is and Challenger was 1 of 25. That's a pretty good jump in reliability over a few years. That was the best we knew how to do in those days and it was plenty good enough. There will never be a totally, fault-free manned transportation system built to go into space.*

Glenn: *We had not been in the program too long when they thought all the problems with the Atlas were worked out. They took us down to see our first launch, which was at night. We were at the camera position and it was a beautiful starlit night. We'd had trouble with going through a max-Q area (maximum dynamic pressure), which is about 37,000 feet, where you're up to a lot of speed but not yet out of the atmosphere. You have the greatest aerodynamic force before you get out into space—that's what Max-Q means. That's where the booster had been failing. They felt they had this whole problem worked out so they had seven new and fresh caught astronauts and they were going to build their confidence by watching this thing go. We're out on the launch pad with the floodlights and countdown begins and the engines lift-off and up it goes. We're standing there watching this thing go up like that and it's beautiful. All at once it hit about 30,000 feet. at max-Q and POW! The whole thing looked like an atomic bomb. We're all standing there looking at each other and it was back to the drawing board.*

Carpenter: *We'd be remiss if we didn't mention Bob Gilruth who started the whole program. He was given the job at the old Space Task Group at Langley (Virginia) of putting together a team. When we went down there I guess there weren't more than probably 25 or 30 total number of engineers in the whole Space Task Group. I remember our first meeting with Bob Gilruth; he called us all in and said he wanted everybody to know since we were experienced test pilots they were looking at us for input to the program. Anytime we felt something was not safe or we were not ready to go and wanted to see something retested, bring it to him and he would take care of it. That was the basis on which we operated. Anytime we wanted to see something done to be safer, Bob guaranteed that was, and if anybody wanted to go back to their parent service at anytime, that was fine. No questions asked. We literally were part of the development team. It's just hard to believe at this point with all of the hundreds and hundreds of engineers and people all over the country involved, from that little humble* beginning under the very fine leadership of a wonderful man, Bob Gilruth, the whole thing started.

Over what piece or part of the procedure was the widest division among you?

Slayton: *I couldn't decide whether we were going to have 12-hour or 24-hour clocks. I remember we decided we'd all try 24. I woke up one morning at the Cape and looked at my watch and I couldn't figure out what time it was. I finally went out to my car and looked at my clock.*

Schirra: *Do you remember a fellow by the name of Harold Johnson who came into our M7 office and gave each of us a 24-hour analog hour watch to wear? We were to wear them to become accustomed to the 24-hour clock because the worldwide tracking network would be using that time during our flights. About three weeks later, Harold came back and one of us asked, "Harold, what time is it?" He looked at the analog watch on one arm, shook his head, pulled up his sleeve on his other arm and looked at his regular watch and said, "It's about 10 minutes until 12." That was the end of that watch.*

Mercury Atlas rocket launched four M7 astronauts into Earth orbit from Pad 14

Shepard: *I think there was division amongst the group about the periscope for latitude control versus the window. We felt that right from the start. We had an air-bearing trainer that was a seat balancing on a column of air, using the periscope. We honestly tried—all of us—tried to make that fly like an airplane, assuming that the gyro system had failed. So, as far as a window was concerned, it was a unanimous opinion, even though it meant extra time and money, there certainly was no disagreement on that. The rudder pedals were something else. I don't know, Deke never could figure out how to fly without them.*

Slayton: *I think what we're talking about was that three-axis stick. The one motion we had to do which was a totally new motion was twisting our hands much like a motorcycle grip for yaw. This would replace the rudder pedals. We'd never flown a three-axis controller before. We'd had plenty of two-axis controllers with pitch and roll, and this was typical in our new fighters. We had to reinvent the wheel, literally, to integrate these three-axis-on-one-hand controllers. I can recall X-15 pilots who couldn't fly our system.*

Shepard and his M7 buddies made their first visit to Huntsville in 1960 to meet Wernher von Braun and check out the Redstone rocket that Shepard was to ride to the edge of space. Von Braun personally took them on a guided tour through his rocket factory, including the Redstone assembly line. The astronauts wanted to get to know the man who was going to provide the rocket power for their ride into space. Shepard was more curious than most. He thought he would be the first to ride the souped-up Redstone that the Army called, "Ole Reliable," the country's most successful ballistic missile. The Redstone was being modified and man-rated to carry the Mercury capsule and Shepard to the edge of space. It was the first time rocket, capsule and man were integrated into a system and launched.

The M7, left, Grissom, Schirra (smoking a cigarette), Shepard, Glenn, Carpenter, Cooper, Slayton visit von Braun and the Redstone.

Rockets and von Braun were something of a mystery to the brotherhood. They understood airplanes and the people who built them. They had flown airplanes in combat and as test pilots. When they strapped on a high performance jet, they knew the machines; and the people who built them. In fact, some actually helped design the aircraft they would eventually control.

This rocket business was different. First, you had von Braun, his Germans and a bunch of young American engineers building rockets in Alabama with little or no knowledge of airplanes. *"Not a stick and rudder man in the crowd, except von Braun himself,"* Shepard would mutter. Von Braun spent a lot of time with the M7, taking great pains to show them how the Redstone was being carefully assembled as the first "man-rated" rocket.

From the blockhouse, the M7 had a chance to watch a static firing of the rocket engine that would power Shepard on the first flight of an American on a rocket. A static test firing refers to a captive firing of rocket engines. The engines are held down in a specially-designed test tower. The idea is to prove the engine works before attaching it to a rocket and shipping it to the Cape for flight. We called it "a poor man's launch." That static firing was one of von Braun's first "fire and smoke shows" and it was spectacular. The fire, noise, vibration and sheer power generated by the Redstone impressed the Mercury astronauts, particularly Shepard. He returned to witness another firing and requested von Braun's permission to stand on top of the test tower, above the engine about where his capsule would be on the real rocket. Shepard wanted to know what it felt like; how much would it vibrate; noise levels; could he read the instruments—things that a true test pilot would want to know before flying any machine.

But there is more to the story. Von Braun did have people walk around the top of a test rocket tower while an engine was being fired to prove to some Congressional types that the rocket was safe. To make it appear that many people were walking within view, he had the four or five people change hats and coats as they disappeared from sight and then returned within view.

Von Braun admired and envied the astronauts. They worked as a team and shared a healthy mutual respect. Likewise the Mercury astronauts admired von Braun, though they were never able to spend as much time with him as they wished. Von Braun devoted most of his attention to the Saturn V moon rocket, while the Mercury guys were busy riding the Atlas and Titan rockets into Earth orbit.

Soon after the Mercury astronauts began training, von Braun entered the picture with a giant Saturn rocket for getting to the moon. The early Mercury and Gemini manned programs expanded to include Apollo. Deke Slayton was calling the shots in the astronaut office….

Slayton: *I think the big breakthrough was getting back to the propulsion side. Wernher developed the Saturn rocket and this was the first rocket designed as a manned machine. That was a big breakthrough. It was also a very high performance machine. In the same timeframe, we were building the command module and lunar module to go to the moon. I don't think any of us, at least I didn't, view the manned part of it as being all that tough of a job. I think we had complete confidence in what we were going to do and how to do it. I didn't have that kind of confidence that they were going to get that big monster together to really make it work until we got pretty far down the stream with it. It turned out to be 100 percent successful and there was never a failure of a Saturn that I know of. Of course, you can attribute that total package to Wernher and his team because that was their responsibility. Someone else was responsible for the command module and the lunar module.*

Glenn: *Wernher and his team had been thinking about manned moon missions for many years. We'd done a lot of studies including a very early program they had called Project Adam. They had thought through, step-by-step, a very crude, early and quick approach to putting man on the moon. This had been proposed in part, when we came into the program. I think their group was pretty much on top of what the man-in-the-loop would do. They certainly did come through, as Deke stated, very well on the booster.*

Schirra: *Without VB we wouldn't have gotten to the moon and back in the time frame we did. He wanted an orbital space station, too. No, doubt he wanted to fly in space. He very much wanted to do that. He certainly sold us on the idea. Shortly after we joined the space program we M7 visited "Huntspatch," that's what we called it in the early days.*

We spent part of the evening in Wernher's home looking through a telescope called the Quest Star…The fun of that evening was when Wernher showed us designs and technical detail drawings of using the V-2 to put man into orbit. These were his great dreams during the 1940s. Here we were in 1959, looking at those drawings. It's amazing how advanced he was in his logic and how far along his team was in planning and thinking about manned spaceflight.

My greatest impressions of VB were two things. One was the way he treated staff and the other was he always had two doors to his office The front door was always open to his subordinates who could get in to see him anytime and the back door was available in case he had to get the hell out of there.

He always complimented members of his staff in front of others, in large meetings where he would bring up a subject and say, "We have this idea, Karl, you tell about it, you thought of it first." Karl was devoted to him from then on. Beautiful technique. Those are the memories I have of von Braun, giving credit to your subordinates for the ideas that they bring up and the front door was always open. That's what changed NASA, they started closing the front door.

My imitations of German accents also caused me to tell stories about Wernher all the time. Whenever we met at some event I had to tell him the story about the cosmonauts and astronauts. At the time, we weren't ready to go to the moon. The fantasy was created where two spacecraft landed simultaneously on the moon, one with CCCP for Russia and the other USA. After landing and the hatches opened, the cosmonauts and astronauts walked toward each other and stuck their bubbled helmets together and said, "Hello Hans, hello Fritz, now we speak German!" Wernher always asked me to tell that story.

I don't know if NASA could handle a von Braun today. They are so bureaucratic. In my day, we had an inspired 'can do' agency. We had a president who was committed. We had Jim Webb who could sweet talk the Hill and the White House and we had von Braun to sell the program.

Carpenter: *In the history of this nation's space program, he's (von Braun) given the short end. I think that's partly because of misplaced patriotism or maybe some professional jealousy. I believe that Wernher's input was irreplaceable to our lunar landing. I think if there are two people to be cited as the prime movers for that program it is Wernher von Braun for the technology and John Kennedy for galvanizing the nation behind the program. Without those two men it wouldn't have been done. Not enough people realize the importance, particularly of von Braun, and in many cases, Kennedy. Without them, we couldn't have done it.*

In a letter dated April 20, 1961, President Kennedy wrote to Vice President Lyndon Johnson, chairman of the Space Council, directing him to take charge and make a "survey of where we stand in space." Johnson asked NASA Administrator Jim Webb, Secretary of Defense Robert McNamara, Dr. Jerome Weisner science advisor to the president and others some key questions about where we stood with the Russians. Of all the people consulted during President Kennedy's mandated space review, no one had been thinking longer about future space travel than Wernher von Braun. He and his associates had been planning space journeys for years. Von Braun was a major contributor to popularizing the idea of manned space flight with a series of articles in *Colliers* magazine in the 1950s, and by collaborating with Walt Disney on space movies.

In his letter to Vice President Lyndon Johnson, von Braun asked to participate in the review as an individual, not as the director of NASA's Marshall Space Flight Center. Von Braun wrote that the U.S. had "an excellent chance" of beating the Russians to a lunar landing. The key questions asked by President Kennedy of Vice President Johnson are as follows:

Do we have a chance of beating the Soviets by putting a lab in space, or by a trip around the moon, or by a rocket to land on the moon or by a rocket to go to the moon and back with a man? Is there any other space program which promises dramatic results in which we could win?

von Braun: *We do not have a good chance of beating the Soviets to a manned laboratory in space. The Russians could place it in orbit this year (1961) while we could establish a somewhat heavier laboratory only after the availability of a reliable Saturn C-1 that is in 1964.*

We have a sporting chance of beating the Soviets to soft landing of a radio transmitting station on the moon. It is hard to say whether this objective is on their program, but as far as the launch rocket is concerned, they could do it at any time. We plan to do it with the Atlas Agena B-boosted Ranger #3 in early 1962.

We have a sporting chance of sending a three-man crew around the moon ahead of the Soviets (1965/66). However, the Soviets could conduct an around the moon voyage earlier if they are ready to waive certain emergency safety features and limit the voyage to one man. My estimate is that they could perform this simplified task in 1962 or 1963.

We have an excellent chance of beating the Soviets to the first landing of a crew on the moon, including return capability, of course. The reason is that a performance jump by a factor of 10 over their present rockets is necessary to accomplish this.

While today we do not have such a rocket, it is unlikely that the Soviets have it. Therefore, we would not have to enter the race toward this obvious next goal in space exploration against hopeless odds favoring the Soviets. With an all-out crash program, I think we could accomplish this objective in 1967/68.

Summing up, I should like to say that in the space race we are competing with a determined opponent whose peacetime economy is on a wartime footing. Most of our procedures are designed for orderly, peacetime conditions. I do not believe that we can win this race unless we take at least some measures, which thus far have been considered acceptable only in times of a national emergency.

On May 25, 1961, President Kennedy made his Urgent National Needs speech to a joint session of Congress, in which he stated, "I believe that this nation should commit itself to achieving the goal, before this decade is out of landing a man on the moon and returning him safely to the earth…"

Buckbee produced television series, featuring Wernher von Braun entitled, "World of Tomorrow"

Von Braun, perhaps more than any other man, has been the driving force behind the moon program. —TIME magazine, July 18, 1969

Sign in the NASA public affairs office during Apollo: "Late to bed, early to rise, work like hell and advertise."

Von Braun's moon rocket was enormous. To work on it, buildings, transporters and other gadgets had to be designed and built from scratch. It weighed in at 3,000 tons, stood 36 stories tall and was a collection of three million parts. To build and perfect it took a unique breed of professionals, a loyal and devoted team of problem solvers. Von Braun's team was made up of young engineers who were doing things for the first time, trial and error leading to cutting edge technology. Working for von Braun was like working for Thomas Edison. He was an inventor of rockets. I witnessed the incredible birth of the moon rocket in his laboratory. It was a place where young engineers realized their good fortune working for the man who was building giant rockets to fly men to the moon.

I was an eager NASA public relations guy who hung onto his every word. I met him for the first time at NASA's Saturn rocket hangar. I was the newest addition to the public affairs office escorting a group of businessmen on a behind-the-scenes tour of the NASA Marshall Space Flight Center. I was to meet von Braun at the entrance to the hangar and according to my watch, he was running late. A NASA government sedan finally approached with someone in the back seat—von Braun. The driver pulled up to the building and stopped; I opened the rear door and von Braun climbed out. I introduced myself and to put me at ease, he acted like we were long-time buddies. I introduced him to each person; he graciously thanked them for coming to the center and said, *"Let me show you the Saturn."* I seemed to have hit my stride as a doorman and held the door open for the group as they filed into the hangar. Once they adjusted to the lighting and looked up at the giant mechanical marvel, they gasped and exclaimed, *"Wow, I can't believe it….Look how big that thing is! How will something that big get off the ground?"* It was a reaction I'd hear over and over, yet I never grew tired of hearing it. As for von Braun, he ate it up. He was in his element showing these executives his creation, the "baby Saturn!" That day, I learned I was in the presence of a unique person who had the charisma, passion and vision to just maybe, pull off this moon landing business.

"Come to America. We are going to the moon!" Von Braun had that one line, attention-getting message cabled to a young German engineer torn between accepting an attractive industry job at home or coming to the U. S. to join the growing von Braun team. He came to America and made a contribution.

I saw von Braun as the consummate communicator and marketer. He could sell his ideas to the man on the street or to members of Congress. Eventually, we became friends and then allies in the dream to build a space science education program and a world-class visitor center to introduce the general public to emerging space technologies. He believed there should be a place where Americans and citizens of other nations could share in the rapidly growing spaceflight experience of the decade.

Managing the affairs of a newsmaker like von Braun in today's world, we would be called handlers or spin doctors, the people who handle public officials and put the spin on the story. We may

Von Braun drove a late model Mercedes-Benz 220 sedan at the time he became director of NASA-Marshall Space Flight Center

have thought we handled von Braun, but looking back, it was more like he handled us. You might assume we had a carefully prepared public relations plan to attract the level of national attention our boss received. We didn't. We were too busy reacting to the demand for his time.

President John F. Kennedy had set the goal of landing a man on the moon before the end of the decade. We were in a race to the moon with the Russians. It was a $20 billion program and we were spending our share. You couldn't walk down the hall of any NASA building without hearing the words, "man, moon," or "decade." We had no idea the man we worked for would make so much history. Our slogan in those days was, *"Late to bed, early to rise, work like hell and advertise."*

Around the office he was referred to simply as, "V-B." In public and in his presence, he was Dr. von Braun. When von Braun visited the White House the Secret Service gave him the code name, "Rocket Man." They used that name anytime he was around the president. I don't believe he ever knew his Secret Service call sign.

Not everyone believed we could meet the President's goal, particularly the national media. They came to Huntsville to see if these big space machines were real and if von Braun was capable of delivering the transportation systems that would safely fly our astronauts to the moon and back. It was a kick to observe the reaction of the media when they walked into the hangar and laid eyes on a Saturn rocket booster for the first time.

His relationship with the media was respectful: He answered their questions without confusing them with technical jargon. The press liked, trusted and believed him. With his handsome appearance and German accent, he became the darling of the space program's press corps. A typical week at the Marshall Space Flight Center in the early 60s could easily include visits of staff writers from *Time* and *Life* magazines, a television crew covering a Walter Cronkite story, science editor Jules Bergman from ABC, and a European television crew or two. Add a couple of science reporters from U.S. newspapers and that would be our press contingent for the week. They wanted three things: a close-up look at the Saturn booster with a chance to kick the tires and take some photographs; to attend a rocket engine firing, preferably a full duration, big booster firing; and an interview opportunity with the man, Wernher von Braun. We made every effort to accommodate the media. Von Braun knew the value of the press in communicating this highly technical program to the public and receiving favorable support from both the public and Congress. He reminded us that the space program was not ours. It was neither NASA's nor was it the exclusive property of the President of the United States and Congress. It belonged to the American people. They deserved to hear how the program was progressing and how we were spending their tax dollars. He would say, *"Americans want to be recognized as leaders in the world, first in technology and certainly first in space exploration."*

In 1964, *U.S. News & World Report* asked von Braun why it was important to go to the moon. He replied, *"I look at the moon as more of a rallying point than an objective in itself. The moon plays the same role in our manned spaceflight program that the city of Paris played in Lindbergh's memorable flight. Paris served as a goal, but surely, Lindbergh had a more important objective in mind than to go to Paris. Our objective is to develop a broad, manned, space-flying capability. You simply cannot develop such a capability...unless you have a clear goal—a focusing point like Lindbergh's Paris—*

something that is universally understood...Now when the President of the United States says, 'Let's land a man on the moon in this decade and bring him back alive,' then you have such a clear goal. Everybody knows what the moon is, everybody knows what this decade is and everybody can tell a live astronaut who returned from the moon."

During the first moon landings, it was customary to hold press conferences at the Cape immediately after the launch. On the platform would be the NASA administrator, the manned space flight director, von Braun and another field center director. Most on the platform were von Braun's superiors. Despite seniority, the majority of the questions were directed to von Braun, especially in the first few press conferences. This embarrassed him. Our solution was to ask some of our friends in the press corps to direct questions to others on the platform, in hopes of getting von Braun's bosses involved. That worked until we encountered a new problem. The foreign press, which was fascinated with von Braun, had the misconception that von Braun was the boss, in charge of NASA. Needless to say, this did not endear von Braun to those at headquarters. In time, von Braun wasn't included in Apollo press conferences at the Cape.

Buckbee confers with von Braun at Saturn V booster rollout and press conference

Often, there would be press conferences to announce a new development in the program. These could become public relations disasters with no real news value. Again, we turned to our friends in the press corps to liven up the press conference. We suggested they ask someone on the platform a question that was certain to get a response. Among the favorites was, "Would the solder joints of the critical components withstand the g-forces and stress of launch?" How solder joints had such a high level of media interest is still a mystery to me. When that question was asked, one of the engineers would give a long dissertation, explaining that these solder joints had been tested to withstand six times the g-forces the rocket would experience at lift-off. The big name science writers in the press corps would get involved in the discussion to show everyone how much they knew about the subject, resulting in a more productive news conference. At a press conference in California, von Braun announced the roll out of the first rocket stage on the production line. After the initial announcement it became very quiet. There were no follow-up questions and people began to disperse. Von Braun passed me a note that read, "I think it's time for the solder joint question." He did have a great sense of humor.

Von Braun had a fascination for machines and man-in-the-loop technology. He wasn't much for robots or unmanned remotely controlled vehicles. He believed humans at the controls enhanced outcomes. He loved to fly aircraft, drive weird vehicles and be challenged by sophisticated simulators. It became well known in the space community if you had a new prototype vehicle that would float, fly or had wheels and you wanted to sell it to NASA, bring it to Huntsville. If von Braun sat in the driver's seat, it was almost a sure thing: NASA would buy it. We went through a period in the 60s when an incredible number of aerospace companies appeared at NASA's entrance with new prototype toys for von Braun to test. He drove moon buggies of all sizes and shapes, flew airplanes that looked too big to be in the sky, and rode on river barges that looked like blimp hangars. I must say, most were successful in selling their machines to NASA because von Braun loved innovation and he had a lot of fun serving as the pilot and test driver.

Von Braun's interest in airplanes began when he became a glider pilot in his late teens. He continued to fly powered aircraft and gliders with the Cumberland Maryland Soaring Club up until his death. He was one of the first NASA center directors—and the envy of many—to get a Gulfstream corporate aircraft. With his love for flying, he made great use of it.

When he took a trip, we all had a regular routine—call flight operations; file a flight plan; collect the speaking materials, Saturn models and giveaways; load the Gulfstream then wheels up in the early afternoon. Once airborne, von Braun would review his speaking points and look over biographical sketches of people who were attending the event so he could call each and their spouse by name. After studying his material, he excused himself and headed for the cockpit. He always asked permission to fly the aircraft, which I thought ironic since it was his airplane and the people who flew it worked for him. Most people at his level on the corporate ladder would have simply taken the controls and flown the aircraft, but not von Braun. He carefully acknowledged people, never taking advantage of his position.

One such flight took us to McAlester, Oklahoma for a speech at the request of Speaker of the U.S. House of Representatives Carl Albert. Von Braun flew the last leg and we landed around 6:00 p.m. at an old World War II airstrip. The official greeting party met us at the airport and we immediately went into a press conference. The local press asked the standard questions in those days, *"How will the space program benefit Oklahoma? Couldn't this money be better spent on some other programs?"* Thirdly, *"What do you think about UFOs?"* He answered the questions as if it were the first time he had ever heard them, with clear and concise answers.

After the press conference, we were loaded into a big limo and joined a speeding caravan, with motorcycle escort and blaring sirens, on our way into the city. Von Braun commented during one of these high-profile visits, *"These high- speed police escorted motorcades are the most hazardous part of my job."* We arrived at the local high school and entered a non-air-conditioned gymnasium, filled to capacity. Von Braun was taken to the speaker's platform for introductions and it was my job to check the visual equipment to support his speech. Standard fare included fried chicken and peas. However, von Braun seldom had a chance to eat due to the constant request for autographs.

From the beginning of his career to the time when he became quite prominent, he never understood the demand for his autograph. Autographs were for rock or movie stars, not rocket scientists. Although he did his duty and signed many autographs, it was obvious when he gave me the nod, he'd had enough and it was time to interrupt and say, *"Sorry, we have a plane to catch."* It was my responsibility to take names and addresses of those who did not get an autograph so one could be sent later.

That evening, von Braun gave the heartland speech, reminding the audience that Americans had always been explorers and pioneers. If Americans want to continue as world leaders, they must establish a leadership role in space. He told a football joke and commended the local congressman for his great leadership in Congress and Washington and his invaluable contributions to the nation's space program. At the close of von Braun's speech, the audience gave him a standing ovation. The congressman, on behalf of the community, named von Braun, "Chief-Fire-Arrows-to-the-Moon," and presented him with a beautiful, full-length Indian headdress. Donning the headdress, von Braun thanked everyone and said his good-byes to the congressman and those on the speaker's platform.

We returned to the airfield, again by high-speed police escort. We climbed aboard the Gulfstream and von Braun went directly to the cockpit and flew us to Huntsville. The professional pilots who flew with him always provided encouragement. He was quick to ask, *"How am I doing?"* The answer, *"You did just fine, Dr. von Braun."* Upon our return from McAlester with von Braun at the controls, we bounced halfway down the runway. As we were leaving, the NASA pilot said, *"You did just fine Dr. von Braun."* After von Braun left and we were unloading the aircraft, I said to the crew, *"It doesn't take a rocket scientist to know that was an awful landing. Why didn't you say something?"* The pilot glanced at me and said, *"Look, I'm not a rocket scientist and you aren't a rocket scientist. He is THE rocket scientist and if he thinks it was a good landing, that's fine with me; subject closed!"*

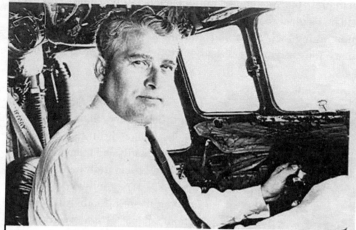

Von Braun who would fly anything that showed up at the airfield is at the controls of an Ford Tri-Motor vintage airplane owned by American Airlines

Von Braun is shown here in the cockpit of a high performance jet fighter

Buckbee was a passenger with von Braun when he piloted this modified Boeing 377 Stratocruiser called the Pregnant Guppy on an early test flight out of Redstone Arsenal, AL. Von Braun contracted with Aero Spacelines to transport his Saturn space hardware coast to coast

Von Braun had some of that showmanship made famous by his friend Walt Disney, the entertainment mogul of the 20th Century. He was literally a salesman of space travel and invited anyone who would listen, to come to Huntsville, the Rocket City, to see a Saturn V ground test. *"It made the loudest sound on Planet Earth. Huntsville became the land of the Earth-shakers, where thunder rolls, smoke billows and for two and a half minutes hell unfolds,"* wrote one Alabama author.

Many came to watch, standing behind a bunker 2,000 yards away from the test stand. As the countdown proceeded for the ground test, water began erupting from the tower, cooling the steel flame deflector to keep it from melting by the 5,000-degree heat the rocket engines created. It would be like a blowtorch cutting through steel if the water were not used to protect and cool the metal.

At first we put special guests in the blockhouse, where they could observe the firing, protected behind portals of several layers of glass and thick concrete walls. Although quite close, they couldn't hear the sound or feel the heat from the test firing. Von Braun didn't like it. He felt the whole point of the demonstration had been lost. He suggested we have them observe from an open bunker from Heimburg Hill. The bunker had a roof, and a concrete wall with a slit for viewing and open in the back. It was sometimes affectionately referred to as the "cow shed." The next time we had a firing, the VIPs were not only impressed, they were startled and some scared out of their wits. These machines were giant powerhouses releasing billowing clouds of smoke skyward with orange and red flames and frightening sounds. The firings became known as, "von Braun's fire and smoke show." It was certainly a way to impress the Washington politicians. Standing 2,000 yards away in a partially protected bunker, they watched a giant rocket ignite, generating 160 million horse power, shaking the ground, feeling the vibration hitting their chest like a hammer and the heat flash directly upon their faces. The first time was scary for everyone. It turned out to be a crowd pleaser that von Braun took great pride in orchestrating.

Von Braun at the controls of Disney World's new people mover

I witnessed over 60 rocket test firings during the Saturn days. On test day, some thrill seekers would elude the guards and maneuver to the rear of an observation bunker, where "they" (we!) couldn't be seen. Standing about a football field away without any protection, we could feel the awesome power of that machine. The heat and the shock waves were frightening. That's when I first experienced rocket engine heat being forced up my pants legs. The sound grew so painful we would put our hands over our ears. No wonder some of us have significant hearing loss. On one occasion I looked over at a friend and he was leaning into the direction of the rocket. He was a little guy at about 110 pounds. The wind and shock waves were pushing him backwards, his clothes were clinging to his body and his hair was blowing straight back. I thought, *"This must be what it's like to witness a nuclear blast."*

I was assigned the job of babysitting a *Life* magazine photographer who came to cover early Saturn V booster firings. He was one of those professionals who had seen and done everything, so he said, anyway. I took him to the test stand to watch the fueling process. Lackluster enthusiasm intact, he asked how close he could get to the test stand during a firing. I drove him to the site which was 2,000 yards away from the stand in an open field with no protection. It was the *Life* photographer, a NASA pickup truck and me. He began setting up his equipment like he was going to take a portrait of a mountain or a river or some landmark. I explained to him that this machine generated a lot of power, vibration, sound and shock waves. In short, he needed to make sure all his equipment was well secured because when this big mother ignites, he'd know it.

Clearly, he didn't appreciate the advice as he explained that he had covered floods, fires, tornadoes and many other natural disasters. This was just another one of those events he would record for posterity; all in a day's work. As the countdown continued, flashing lights and sirens began sounding, warning people in the area that a firing was imminent. The countdown hit zero and just for a moment, there was complete silence. Then a giant gulp of burning kerosene belched from the mouth of the test stand. Suddenly, roaring sounds were everywhere, trees were blown, the ground shook and heat reddened our faces. I was totally captivated by the firing; my eyes glued to the flaming monster. A movement from the corner of my eye distracted me—the *Life* photographer was gone. He was running across the field, away from the firing. When he left, he knocked over the tripod and camera and never captured one picture. Later, I made arrangements for him to get a negative from one of the NASA photographers who had captured the firing from another site not far away.

In addition to popular media we had our share of White House VIPs; Eisenhower, Humphrey, Kennedy, Lyndon and Lady Bird Johnson and Ford.

A call from the White House came to our public affairs office in February 1964, inquiring about supporting a visit of the first lady, Lady Bird Johnson, to Huntsville and the Marshall Center. This was not only a visit by the President's wife, but as important, it was to be Lady Bird's Alabama family reunion. Lady Bird's mother was born in Alabama and she had many kinfolk living in the state.

She was described to us as a human dynamo with a pleasing personality and the mind of a Wall Street banker. She read by an oil lamp during elementary school and had few playmates. One of her brothers gave testament to the lonely setting of her childhood home when he said of their small town, *"The rooster crowed only once, for he heard no answering call."*

She arrived on March 24, greeted by Dr. and Mrs. von Braun, Mr. and Mrs. James Webb and a score of local officials. Following the first lady was the White House press corps and other representatives of the fourth estate from the Southeast, the "unwashed herd" we had encountered on previous White House visits.

Lady Bird greeted 60 of her favorite Alabama relatives who arrived by private jet, commercial airlines, limousine, private car, train and a horse. A guy from Monrovia, Alabama wanted to demonstrate he was a true cowboy, decked out in the hat, cowboy outfit and boots. Needless to say, he received a lot of attention from the security people at the entrance to Redstone Arsenal.

Wearing the cowboy hat given to him by President Johnson, von Braun presents first lady Ladybird Johnson a hard hat while visiting Huntsville

Von Braun donned his authentic Texas hat given to him by President Johnson at the LBJ Ranch several weeks before, and presented the first lady with an official NASA hard hat personalized with her name. It was a great photo opportunity with America's first lady from Texas standing next to "Rocket Man" Wernher von Braun.

The first lady was one the few people in the world to observe back-to-back static firings, a proud moment for test director Karl Heimberg. He conducted a significant first, two tests, one of a Saturn booster followed by a second test of an F-1, on the same test stand with only ten minutes between the two firings. It was an impressive show and Lady Bird was deeply moved.

In her remarks later that day, she said, *"There are many reasons I have looked forward to this visit today, to this great space flight center. Not the least of these is the fact that Lyndon has, since 1958 when he authored the Space Act, made the establishment of our national program for peaceful exploration of space one of his major thrusts. The depth and intensity of his interest has, very naturally, stimulated my own, and I welcome this opportunity to see how our objectives to use space for peaceful purposes are being met. I am particularly pleased that my own personal exploration of the space program should begin here in Alabama—a part of the world that is almost second home to me. I am very proud of the vital role which Alabama and the South are playing in our national space program. And you who have taken such a big hand in it must be very proud. As I stand here today and gaze out over the rolling lands covered with pine trees, I am mindful how much the story of our country's past and future is written in the red clay hills. On these very lands, now dotted by rockets...cotton once was king. Today, we look not to the past, but to the future. For the South has hitched its wagon to the stars."*

The first lady sent a personal thank you note to Dr. von Braun, which stated in part, *"Stimulating, exhilarating, informative—these words cannot fully describe my thoughts concerning my visit to the Marshall Space Flight Center on Tuesday. The significance of the Saturn has come home to me."*

Wernher von Braun always looked to the future. He looked upon today's children as tomorrow's explorers, the ones who would keep the dream alive for expeditions to the moon, to Mars and other beckoning worlds of the solar system. He often said that we should not think in terms of building for today; rather, we should invest in the future.

Von Braun believed it was an obligation to help prepare young people by giving them a glimpse of the future. It was his idea to build a Huntsville space science center. He believed there should be a place where Americans, and indeed, people from all over the world, see our accomplishments in space. He believed it should be in Huntsville, Alabama because that's where the space program began. He lobbied the State, developed a plan and assigned me to work on his ideas for a showcase of rocketry, space exploration and discovery. The result of his creativity and salesmanship is the U.S. Space and Rocket Center and U.S. Space Camp in Huntsville, Alabama.

Von Braun was fortunate to see many of his dreams come true, but there was one that never came to fruition. It was his desire to fly in space, to experience zero-gravity and to re-enter the earth's atmosphere at 15,000 mph. When he was around astronauts like Alan Shepard, John Glenn, Wally Schirra and Deke Slayton, he always had questions—What did it feel like at lift-off? Was it easier to work in zero gravity? Could you see anything during the fiery re-entry?

During an interview in 1972, Von Braun was asked would he like to go and he commented, *" I'd love to."* He explained that it would not be possible in the moon-landing program because Apollo astronauts were professional astronauts specially trained for the moon landings. However, in the space shuttle, in addition to the pilot-astronaut, scientists would be accommodated. Asked if he had bought his ticket, Von Braun replied, *"I hope they give me one or I could thumb a ride at a reduced fare."*

Being a pilot, he understood space flight was the next step in the challenge. I think he planned to fly as a scientist aboard the space shuttle, so it could be deemed safe for other non-astronauts and even politicians. The fact that he, a non-astronaut, could safely fly in space, would set the stage for the President of the United States—the leader of the Free World —to address the world population from space. It would also help him and the team to develop the next generation space shuttle. Unfortunately, events did not unfold in time and von Braun was taken from us in 1977 at the relatively young age of 65.

In August 1969, a few days after Neil Armstrong walked on the moon, von Braun made a presentation in Washington that contained a detailed plan to land man on Mars. Unfortunately, the space program slowed down and virtually went into a holding pattern.

In 1975 von Braun wrote for the *Washington Star*, *"As one gets older and approaches retirement age, somebody else picks up where you left off. There were great men long before the first big rockets were built. We were building on their legacy. We want to make sure this legacy can now be passed on to the next generation, the people who will really pick the fruits of trees we have planted. I think the silliest part of the decay of the public interest in space is that we planted the orchard and we nourished it, fertilized it, and watered and gave it our entire tender loving care. And now, when the fruits can be picked, they don't want to pay the fruit picker. This is where I think the younger generation can make the greatest contribution. Pick the fruit!"*

One of von Braun's NASA Headquarters' bosses, Alan Lovelace stated, *"In the tradition of Newton and Einstein, he was a dreamer pursuing visions, and at the same time a creative genius. He was a 20th century Columbus who pushed back the new frontiers of outer space with efforts that enabled his adopted country to achieve a pre-eminence in space exploration."*

Fame and accolades came late to von Braun. It was not until the 100th anniversary of human flight in 2003 that he received what I believe to be his highest honor. He was named the second most important aviation pioneer in world history, second only to the Wright brothers. Nine others who are named to the list of 100 top aviation pioneers achieved that honor by having ridden or flown experiments on rockets that von Braun conceived or developed. They were astronauts Armstrong, Aldrin, Shepard, Grissom, Schirra, Lovell, Young, Ride, Glenn and scientist Dr. James Van Allen. Unfortunately, von Braun didn't live to receive the honor but he would have been pleased to be in the company of so many of the brotherhood.

I still marvel at his vision and sense of purpose. He not only dreamed about space flight, he set bold goals and objectives. He molded the team that put the U.S. in space and enabled America to be the first and only nation to walk on the moon. He not only believed it could be done, he taught us how to do it. He created a new breed of thinkers and problem solvers. He helped us touch the stars and left behind a blueprint for exploring the universe. He was a crusader for space travel. We miss him. Space exploration misses him. If he were around today, he would give his heartland speech, the one that calls for this country to maintain a leadership role in space exploration and push cutting-edge technology. Then he would challenge us, as Americans, to set bold and difficult goals. He would say, *"It's time for an American to take another walk and this time on Mars."*

Buckbee and von Braun view first exhibit for the Huntsville Space & Rocket Center

Walt Disney and von Braun, two visionaries from different worlds, worked together on many projects

Buckbee and von Braun inspect rockets on
display at the Space Center.

One lesson I learned early in my public relations career, never mention monkeys, chimpanzees or primates in the same sentence with astronauts. If the Mercury astronauts detested anything, it was being compared to the chimps that flew before them. No Mercury astronaut was ever heard saying anything complimentary about chimps or monkeys. In fact, there was more than one Mercury astronaut who believed we would have beaten the Russians and claimed space first had it not been for so many chimp flights.

The Mercury astronauts had to put up with engineers and project managers who believed the chimps could achieve results more cheaply and faster without risk of human life. Of even more concern to the brotherhood, was upon returning from a space flight, the astronauts' reaction times were often compared with the chimps'. This was not well received, especially by Shepard.

One male chimp named Enos, a favorite of the engineers and medical cult, was often displayed in front of the press upon returning from a flight. Each time his handlers brought him before the press, he would rip off his diaper and display an erection. This caused members of the press corps, especially the females to comment, *"Why doesn't NASA do something about that monkey?"* Even after considerable editing, Shepard's comments about Enos were not printable.

Chuck Yeager and other brethren, the Air Force sound barrier-breakers, had no mercy. Yeager referred to Shepard, the M7, and their capsule as, *"Spam in a can."* Yeager scoffed, *"There won't be any flying to do—a monkey's going to make the first flight."*

Buckbee and Shepard during 20th anniversary of the first moon landing

I have yet to meet an astronaut who had any love for these apes. Whenever I worked on public appearances, speeches, publications, exhibits or broadcast interviews, it was forbidden to discuss astronauts and chimpanzees in the same breath or sentence. I asked Shepard about this and others who appeared to have hang-ups regarding the apes. He said, *"I get a little tired of checking every simulator and capsule seat for primate poop before I climb in. I didn't know what a primate was until I got into the astronaut program. Right away people began comparing my reaction time to these damn monkeys. I didn't sign up for this program to be a freaking specimen or test subject. In Hanger S at the Cape where our quarters were, we could hear that small colony of apes howling and screeching every time we walked down the hallway."*

I reminded Shepard I heard from some NASA engineers that the reason he got to fly on a Redstone was because NASA ran out of monkeys. In fact, primate flights did delay plans to fly both Shepard and Glenn when engineers wanted one more test flight before strapping the Mercury guys into the capsule.

Later, I became the custodian of my own space monkey, Monkeynaut Baker. I invited Shepard to attend a celebration honoring Baker, a squirrel monkey who had successfully flown before Shepard aboard a Jupiter rocket. I thought it would make a nice photograph; America's space pioneers, Monkeynaut Baker, one of the only living monkeys to have flown to the edge of space, and Alan B.

Shepard, Jr., America's first astronaut. I received a rather nasty reply from Shepard, so needless to say, I didn't get the photograph.

Monkeys Able and Baker were the first primates the U.S. flew on a Jupiter rocket in 1959. Von Braun and the U.S. Navy teamed up to develop the spacecraft and the rocket. It was a successful launch and recovery. However, Able died shortly after recovery from complications caused by an injection given to him to remove sensors from his body.

Miss Baker had made headlines on May 28, 1959, when she and monkey Able, her traveling companion, became the first American animals to enter space and return to Earth. The two primates flew in a specially-developed, semi-closed life support system—fitted into the nosecone of a Jupiter intermediate range, ballistic missile, a product of the von Braun rocket team while working for the Army.

Monkeynaut Baker, the first lady of space

Able and Baker spent nine minutes in space, reaching an altitude of 300 miles, traveling at 10,000 miles per hour. Their nosecone splashed into the Atlantic Ocean some 1,700 miles downrange from the launch pad at Cape Canaveral. Baker made the sub-orbital flight wearing a helmet and reclining on a rubber pad with her knees drawn up. Her couch was positioned in the nosecone so she was facing the direction of flight. Thus, the force of launch and reentry was directed from Baker's chest to back, just as it would be with human astronauts like Shepard who would follow her into space. Special bio-sensors attached to Miss Baker's body measured her respiration rate, body temperature and pulse. These physiological parameters, as well as the cabin pressure, were telemetered to Earth throughout the flight and after splashdown.

Following her missions, Miss Baker returned as a celebrity to her former home at the U.S. Navy Medical Research Center in Pensacola, Florida. There, she was placed in an elaborate room with Formica walls and ceiling, a tile floor and air-conditioning. It was specially lighted and had a one-way window so that scientists and visitors could see in, but Miss Baker could not see out.

Once the Navy medical personnel were assured they had gained all the information Miss Baker had to offer on the effects of her space flight, they began seeking a permanent home for her. Rumor was they were expecting Baker to pass away and I discussed the idea of bringing her to the Space and Rocket Center with my deputy director, J. Scott Osborne. A creative exhibit designer, Osborne designed a "monkeynaut habitat" for Baker to live out her remaining years.

"The Navy maintained and observed Miss Baker during the 10 years following her historic 1959 flight. The problem was, she was expensive to care for, so, with impending budget cuts, they let it be known they were accepting proposals from various organizations to take over the care of Baker," noted Osborne. The Washington Zoo, San Diego Zoo, Kennedy Space Center and Huntsville's Space and Rocket Center were interested.

To add support to our bid, I asked von Braun to help and he agreed to sign a letter requesting Baker. The letter and proposal explained how we planned to display the animal, how we planned to present her in the manned space flight story and display the rocket developed in Huntsville that propelled her voyage.

Osborne went to Pensacola to inspect Baker's habitat and also consulted with veterinarians at the Yerkes Primate Center in Atlanta, Georgia.

In our proposal, we agreed to build Miss Baker a special, climate-controlled habitat, providing full-time staffing to tend her, including her own personal vet—all expensive propositions. It was an elaborate proposal including a closed-loop environmental system for the animal. She would live in her own air-conditioned system. The Navy liked the proposal, but apparently there was some pressure by a politician to have the animal sent to the Washington Zoo. I advised von Braun of this and he made a personal call to the chief scientist at Navy Pensacola. Soon afterwards, the animal was flown to Huntsville in a Navy plane with a veterinarian and staff. We were awarded the animal because of Osborne's design and creative proposal, and of course, von Braun's interest and influence. He never failed to support one of our projects if it contributed to Huntsville's manned space flight story.

Miss Baker's new habitat chamber had to be cleaned daily, plus contingency plans had to be made in the event of power outages. Special lighting to match daylight, constant temperature and humidity, fresh air from the outside of the center (to lessen the risk of human diseases) and constantly running drinking water were but a few of the accommodations required for housing Miss Baker. In addition to her not-so-private quarters, Miss Baker had a personal handler, David Gaines, and veterinarian, Dr. Charles Horton.

Baker received hundreds of fan letters per week. With the help of Doris Hunter, she responded with a personalized answer which typically read:

Thank you for your recent letter. It was very nice of you to write to me and I will try to tell you about my space flight and myself.

I was born in the jungles of Peru. When I was very young I was brought to the United States and selected for space flight training. A number of monkeys were tested in simulated space capsule conditions. I passed all the tests and was selected, along with monkey Able, for a spaceflight to determine if it would be feasible for a man to journey into space. Able and I were clothed in tiny space suits and helmets, just like the astronauts.

Monkeynaut Able was a Rhesus monkey. He was born in the United States, and at the time of our flight weighed 7 pounds. Able did not live to be old, but died shortly after we completed our space flight.

After the flight with Able, which was the only one planned for me, I retired to live with my husband Big George. As you probably know, in 1971 we moved to Huntsville, Alabama, and since that time our home has been at the world's largest space museum—the Alabama Space & Rocket Center. I must tell you that Big George died quietly in his sleep on January 7, 1979. He was 17 years old.

Because it is not good for me to live alone, I was married on April 1, 1979, to Norman from the Yerkes Primate Center, Atlanta, Georgia. He is a very nice monkey and very handsome. We will continue to live at the Alabama Space & Rocket Center in my specially designed house that is called the Monkeynaut Chamber. It has its own heating, air conditioning and water systems, and is a very comfortable home.

My days begin at 8:00 AM when I get up and have my daily meal. I eat parts of a banana and orange; hard-boiled eggs, some strawberry Jell-O with extra vitamins and monkey biscuits. Then, I relax for a while and usually have a nap before the visitors start coming by. I spend a lot of time watching them. I exercise on my bars and swing, and I love to do tricks for the children who stop by to watch me and say hello. I retire for the day and go to sleep at 8:00 PM. So you see, I have a very full and interesting life.

I enjoy living at the Space & Rocket Center. It's an exciting place that has the world's largest collection of rockets and space vehicles. You can operate many of the exhibits yourself and pretend that you, too, are an astronaut. I am honored to be a part of this great exhibit that features our nation's accomplishments in space. Everyone is very good to me. Often, after the Space & Rocket Center closes for the day, Mr. Gaines, the nice man who takes care of me, gives me a ride around the Space Center in a baby stroller. It is very exciting and I get to see all the new and different exhibits that are being added all the time.

Again, thank you for you letter, and I have enclosed a picture of Norman and me that you may keep. Come to visit me at the Alabama Space & Rocket Center real soon—I'll be looking for you.

It was signed with a paw print of Monkeynaut Baker.

Not only did Miss Baker entertain millions of visitors during her 13 years at the Earth's largest space museum, she also was familiar to millions of TV viewers across the nation. I traveled with Baker and her entourage to appear live on the *Dinah Shore* and *Mike Douglas* shows. Miss Baker and I shared the stage with George Carlin, James Brolin, Sandy Duncan and Joyce Dewitt.

Temperature and humidity were always problems when traveling with Baker. United Airlines, the "official carrier of Miss Baker," normally upgraded us to first-class because the captain could better control the temperature in the smaller cabin. Baker had a seat because of her reputation as being the first lady in space (sometimes to the dismay of other first-class passengers). At hotels, we placed her in the bathroom and ran a steamy, steady shower to keep her comfortable.

Birthdays were frequently an international media event. At the time of Baker's 25th birthday, Dr Horton was contacted by the *London Daily News*. The reporter wanted to know if Baker's longevity was due to her space travel. Dr. Horton noted that was not the case and added factors such as her controlled environment and indomitable spirit were probably the reasons. Time passed and he received another call, this time from the BBC (British Broadcasting Corporation). The same questions were asked and Dr. Horton gave the same answers. Two weeks later a letter addressed to Dr. Horton, lambasted him for cruelty to animals. Something had been lost in translation as this animal lover thought Miss Baker had been flying around in space since May 1959!

Schirra visits Miss Baker, first lady in space, on her 25th birthday. Shepard was invited but declined to attend

Neil Armstrong, Apollo 11 crew and first man to walk on the moon, sends Baker a birthday message from an admirer

Gene Cernan, last man to walk on the moon, sends his congratulation to the first lady of space on her silver birthday party celebration

Miss Baker passed away on November 29, 1984. She was buried at the entrance to the U.S. Space & Rocket Center with a stone marker:

Miss Baker, Squirrel Monkey
Born 1957, Died November 29, 1984
First U.S. Animal to Fly in Space and Return Alive
May 28, 1959

Since her passing, Monkeynaut Baker has been recognized by several different organizations for her early achievements and contributions to space flight. In a black tie event held on January 29, 2005 at her final resting-place, the U.S. Space & Rocket Center, Baker was inducted into the Alabama Animal Hall of Fame.

Shepard was proud of the fact that he was not only the first American astronaut to ride a rocket, but also was the only space flier to ride von Braun's Redstone and Saturn V moon rockets. He was a bit curious about the Redstone and often asked me what the von Braun team did to the Redstone missile to make it man-rated for him. My stock answer was, *"We added more fuel and increased the amount of explosives in the destruction package in case it went off course and had to be destroyed."* Actually, I wasn't sure what was done until I found a report in NASA's history office written by Jack Kuettner, von Braun's project manager for the Mercury Redstone program. Written in 1958, it was entitled, *The Launching of a Manned Missile*. It must have been well done because von Braun had added a handwritten note, "Very fine paper!" and signed it with his famous "B."

In 1959, left, Shepard, Slayton, Grissom, von Braun, Cooper, Schirra, Glenn and Carpenter inspect Redstone hardware with Project Manager Jack Kuettner, right in the rocket factory in Huntsville.

I showed it to Shepard in 1989 when we were collecting artifacts and memorabilia for the Astronaut Hall of Fame. He had never seen or heard of it. The following are excerpts from the paper and some interesting responses by Shepard:

> In the launching of a manned missile, conflicts are encountered between aircraft technology and missile technology. These stem from the difference between man-controlled and automatically controlled vehicles. The possible roles of the occupant as a living payload, as a manned backup for automatic controls, or as a true pilot during the powered flight are considered. Reliability standards of missiles versus manned vehicles create problems of pilot safely. Protection of the occupant on the launching pad, after firing

command, then during the rest of powered flight must be made compatible with existing launching procedures. Protection of life other than the occupant's must also be taken into consideration. A trend away from the aerodynamic instability of boosters is hopefully expected. Automatic escape systems will have to rely on a small number of sensors covering a wide range of possible emergencies. To what extent redundancy can and should be achieved needs careful scrutiny. Finally, the possibility is considered of training space travelers through simulated rides during static firings.

By putting man in a missile, the separate avenues on which these technologies have proceeded are made to intersect, bringing into sharp focus the following questions: (1) What is the role of the occupant? (2) To what degree is the safety of the occupant assured? (3) What special developments and procedures are required to integrate man into the missile system?

The role of the occupant is, of course, that of a living payload which, unfortunately, is very sensitive and positively indispensable. As such, missile engineers eye him with misgivings. In turn, existing boosters are, in the eyes of the occupant, barbaric machines grazing human tolerance limits in all directions. In this cargo function, the occupant is strictly a problem child.

Watching Shepard's reaction while reading the above part about eyeing the occupant with misgivings and being strictly a problem child, brought on the comment, *"What the — is he talking about?! Barbaric machines, occupants, problem child; he's talking about me! Who is this guy? I think I met him one night at Wernher's house. Wasn't he a fighter pilot in the Luftwaffe?"*

I reminded him it was written forty years ago and, more importantly, before a manned flight on a rocket. *"Read on, José, it gets better,"* I said. He did like the part about training the occupant through simulated rides during static firings.

Kuettner continues:

More justice and more satisfaction would be given to the human occupant if he were used at least as a backup for the automatic controls. Taking missiles as they presently are and coming down to facts, we discover that there is little time for intelligent pondering during the powered phase and that therefore, a detailed information display is of little help to the occupant. Man may be a lightweight computer of amazing versatility, but in missile and computer terms, he is outrageously slow.

This brought Shepard out of his chair. *"Look, we were all trained as test pilots. In aircraft, we were expected to be the primary system and we proved over and over again we could react and save the aircraft. Why should it be any different with rockets? If the engineers give us the right tools and controls, we can be more than an 'occupant.' I hate that word and I'm not real fond of 'backup' either. We astronauts should always be considered the primary system when designing a system like an aircraft or rocket. As to his comment about being 'outrageously slow,' well, that sounds like an engineer talking. We had many heated discussions with engineers about controlling the spacecraft and this was after Mercury Redstone. Remember Gordo's last Mercury flight. He had a total power failure, we are talking no onboard power. He came close to having a fire. He took over manual control and lined the spacecraft up for reentry. There he was, using his piloting skills, his eyeballs and manually brought that spacecraft within the footprint of the carrier, without a computer, I might add."*

Shepard continued, *"At the beginning of Gemini, we got our control in the Gemini spacecraft— over a lot of dead bodies in the ranks of the engineers. That's why Neil (Armstrong) was able to save Gemini 8 when it went out of control. He took over manual control and saved the crew and the mission. Gemini was probably the best designed spacecraft for flying in space that an astronaut would ever want to strap on and that's because we astronauts were involved in the design from the beginning."*

Kuettner's report labored on:
> Man's value as an intelligent being would come fully into play if he were at the controls of the missiles in the true sense, determining flight path, thrust, etc., as we are accustomed to seeing in space fiction stories. For existing hardware, such a function is completely out of the question. Whether or not this will change in the future will depend on whether the booster and manned payload can be designed as an integrated system. It is likely that the required large boosters will be used as versatile work horses designed to carry manned, as well as unmanned upper stages, and will therefore have to incorporate automatic controls. The X-15 flights on the one side and the Mercury flights on the other side will give the first answers to this question. After separation from the booster, things may be quite different.

"The X-15," Shepard commented, *"certainly gave us an indication of a pilot's value in being the primary system in a highly sophisticated rocket plane. Look at what Scott Crossfield and Neil Armstrong were able to do with that machine that featured rocket-powered engines, guidance and control with thrusters, and fly-by wire. It didn't fly to the edge of space by itself."*

Continuing the subject, Kuettner wrote:
> In introducing human life into missile technology, certain missile standards change entirely. What has been considered a most successful missile may turn out to be a hazardous monster in terms of conventional flight testing. Let us look first at the reliability angle, being careful to distinguish between mission reliability and survival reliability. As manned flights should be made only with well-tested missiles, one can base the reliability estimates on actual firing records, rather than on a paper prediction according to components. We should clearly state that we could be only 90 percent sure that the mission will be accomplished in 7 out of 10 cases. Or, to make it sound a little better, we may say that there is only a 5 percent chance that 3 out of 10 missions will fail, and there is a 5 percent chance that 9 out of 10 missions will be successful.

Never at a loss for words, Shepard noted, *"Well with those odds, I know some guys who might not have taken a ride on those rockets."*

Back to the report:
> Although this is a somewhat frightening situation from an aviation standpoint, one must realize that the survival probability would be much higher than the mission reliability. In this case, the confidence limits will be quite low for lack of sufficiently large test samples. However, pioneers have taken much greater chances in the past. Due to the nature of their undertaking, the chances have been generally unpredictable.

"Who said we weren't pioneers ourselves," quizzed Shepard.

Kuettner's report:
> Certain hazards must be considered from the moment the occupant boards the missile. Whether he likes it or not, he will probably go aboard more than an hour before firing. He will be placed in a sealed cabin on the top of a large, fueled missile surrounded by a whole array of solid rocket systems, high-pressure bottles, H_2 O_2 containers and electrical sub-systems. Once the service structure has been removed, is the occupant sitting on a volcano ready to erupt? For well-tested missiles, the acquired experience in firing and count-down technology definitely suggests the answer: No. Nevertheless, escape methods must be

provided, even for the remote probability of a pre-launch mishap. Ejection is not the answer to all emergencies. It will subject the occupant to tremendous accelerations, as it is designed to save him from a possible fireball. It requires the functioning of a complicated sequence of events; it may create a hard impact with possible damage to the cabin; it may not save the occupant from an internal cabin emergency. The occupant should be able to leave the cabin in an emergency after the service structure has been removed.

As soon as the firing command is given, another group of possible emergencies must be considered. If, for example, the thrust is insufficient to lift the missile, the engine may be cut off, and no damage is done. In case of tail fire, or if the missile lifts off only to fall back, quick ejection is imperative. The same is true if the missile goes out of control shortly after lift-off. How shall the ejection be activated? By the occupant? By the firing crew in the blockhouse? By the range safety officer? By automatic sensors? Shall the engine be cut off? In reaching a solution, there are several aspects to be considered: (1) What is the time factor? (2) Who has the best insight into the situation? (3) Is human life other than the occupant in danger?

The time factor may exclude all but automatic ejection. The occupant has probably the least insight into the situation of all parties involved, since the missile is under exhaustive ground surveillance by telemetry. From a standpoint of danger to human life other than that of the occupant, the range safety officer may exclude premature engine cutoff. Consequently, automatic ejection without engine cutoff may be a compromise. Backup by ground command and by manual pilot control is desirable.

In an abort system, the space capsule itself will, of course, have to sense its own internal malfunctions, but here the occupant plays a more active role. Three questions arise in the design of an abort system: (1) Shall the system be located in the booster or in the space capsule? (2) Should it be made redundant? (3) What information shall be displayed to the occupant in order to use him as a backup?

Since the malfunctions under question originate in the booster and since manned payloads may be fired on different types of boosters, it appears naturally to locate the abort systems in the booster. As to the last question, the analysis shows that the emergency cases sensed by the automatic systems are of such critical urgency that even the most intelligent occupant will have a hard time trying to interpret displayed information. There is little time to tell him the cause of the emergency; he will just have to push the ejection button if the abort signal comes on and the automatic systems fails. This cannot be helped. Of course, the normal capsule instrumentation will give the occupant sufficient information on the general course of events.

After digesting the information, Shepard responded, *"Well, if this had been the case, Skyray (Wally Schirra) wouldn't be with us today. If you remember on his Gemini-Titan launch, he had an automatic shutdown on the pad and didn't hit the chicken switch (ejection controls) because he knew they didn't have lift-off. You know why he knew they didn't have lift-off? Because he couldn't feel it in his butt, in the seat of his pants. How did he know that? Because of his previous experience in his Mercury flight. My point is the engineers have to give us some credit for having the experience and skills to make decisions during such critical times. If they leave everything up to the machine, why even put 'occupants' on board?"*

Back to the report:

Space-medical training will take care of most of the problems to be faced by the astronauts during the space ride, except of course, prolonged weightlessness. Some special thought has been given to the time before and after fire command, when the space traveler must sit on top of a huge fueled missile and wait for the rocket engines to fire, exposing him to powerful noise and vibration. How can he be prepared for this experience? Is it feasible and safe to give the passenger a simulated ride on an actual missile during a static firing?

Shepard read this and said, *"Well, that's exactly what I proposed to Wernher in 1960, that we all be permitted to stand atop the Redstone tower during ground tests."*

"Did he let you do it?" I asked.

"Hell no, he didn't let me do it. Later, I heard he let some of his engineers and test conductors walk around the top of the tower to show off in front of some Washington politicians. I don't know what that proved," grumbled Shepard

Winding up his report, Kuettner finished:

For a well-tested engine on a fully equipped test tower, the answer is definitely, yes. Reliability studies show that we are well above the 99 percent level and those existing and added safety devices can take care of practically every situation, such as tail fires. Experience indicates that there is plenty of time to use the connecting tunnel (with the test tower and bunker), in both directions in case of an emergency, but the need will, in all likelihood, not arise.

No manned missile has as yet been fired. We are just feeling our way into an unknown terrain. The integration of man and rocket will, for sure, take its course until a clear pattern develops. The ideas discussed here originated in the minds of many members of the von Braun team at Huntsville, Alabama. These ideas have their basis in long experience and are necessary because the Army Ballistic Missile Agency is actually faced with the problem of launching the first man into space in the near future.

" And that was me, May 5, 1961," Shepard added.

CHAPTER 7
Launch of America's First Astronaut:
Roger, lift-off the clock has started…
"You are on your way, José!"

Our opponents across the ocean, behind the Iron Curtain, thought about a month ago they had slammed the door to the universe in our face, but Shepard has let us out of our dilemma and embarrassment. We will go farther and farther, eventually landing on the moon. —Wernher von Braun, May 5, 1961, Huntsville Alabama

It was my first trip to Cape Canaveral. I was offered a tour of the Redstone rocket launch pad where Alan Shepard or John Glenn would make history. No one was sure which one because NASA was keeping it a secret right up until launch. The Army guy taking me around asked if I would like to come out and "watch them blow up this crazy astronaut." That seemed to be the tone at the Cape. Malfunction Junction, as the Cape was jokingly referred to, had seen more than one mishap. A lot of rockets were blowing up, some in the presence of the M7 astronauts. There had been at least two scrubs of Shepard's launch. But now it was show time and I was fortunate to be present for the main event on the morning of May 5, 1961—the launch of America's first astronaut, Alan B. Shepard Jr. aboard Freedom 7.

I was there to survey government property believed to have been destroyed by an earlier rocket launch. The launch, as it was told to me, began with a magnificent lift-off, but instead of going ballistic like most rockets, it immediately turned right and flew horizontally in a westerly direction, straight toward downtown Orlando. The range safety officer intentionally destroyed it. Eberhard Rees, von Braun's longtime technical deputy, when asked what was learned from this failure, said, *"The destruction package worked perfectly."*

Interestingly enough, there was a lot of government property allegedly destroyed by this malfunctioning rocket according to the documents I was to verify. Pick-up trucks, trailers, tape recorders, cameras, sophisticated test equipment—all government property not normally found on a rocket launch pad—came up missing. I figured every piece of government property reported missing at Cape Canaveral in the last three years had been on or near that launch pad.

I was standing with a couple of Army photographers about a mile away in an unprotected area that was heavily infested with snakes and alligators. There were no viewing stands or special VIP area. We waited through several delays before it finally got interesting. The countdown—we could barely hear the garbled launch director over a static-ridden field telephone—was at one minute and counting. As zero approached, fire and smoke billowed from the rocket, an arm dropped away, and bang! It was off the pad and moving skyward. Incapable of speech, we watched with hearts pounding, knowing that rocket was taking us all someplace, but we had no idea where. We held our breath, expecting at any moment it would either blow up or turn hard right and end up in Orlando. The rocket flew out of sight into the clouds and disappeared. Little did I know that years later I'd be hanging out with America's first astronaut, Alan Shepard.

In 1991, on the thirtieth anniversary of Shepard's flight, we reminisced about his first historic moment from Pad 5 at Cape Canaveral. We were preparing answers to questions from the press and laying out an exhibit storyboard. I asked him why we were second to the Russians. Shepard replied, *"Because von Braun and you guys in Huntsville were too damn conservative. You wanted another unmanned flight after the monkey screwed up. Well, actually the monkey didn't screw up; it was some little electrical relay. We had the Russians by the short hairs and we gave it away."*

I reminded him it was reported that he and the whole Mercury team were asleep at the time. Is that how you heard about the Gagarin flight? *"Yes,"* Shepard answered, *"I got a call from a friend and I was asleep. He told me the Russians had done it. They had put a man in orbit. And I thought to myself, I could have been up there three weeks ago."* It was a bad time for the astronauts and it was a shock

Russia's Yuri Gagarin becomes the first human in space, 23 days before Shepard

Chimpanzee Ham suited up for his first flight in 1961 aboard a Mercury Redstone

to the American public.

Colonel John "Shorty" Powers, the spokesman for M7—chose to make an unorthodox debut on the space fliers scene. He had a bit of a swagger and liked to be thought of as the eighth Mercury astronaut. He also was asleep but in this case was awakened by a needling reporter who wanted to know what was going on at the Cape. Shorty shouted, *"If you want anything from us, you jerk, the answer is we're all asleep."* The morning headline read, "SOVIETS SEND MAN INTO SPACE; SPOKESMAN SAYS U.S. ASLEEP." So much for the Mercury spin-doctor.

Shepard continued to reminisce. *"We were training at the Space Task Group at Langley Air Force Base in Virginia. Bob Gilruth, who was the director of the Mercury program, called us into his one-room office. It was very quiet. Nobody said anything at first."*

Gilruth was just about to tell seven top-notch, competitive, highly motivated pilots, that one among them had been picked as the first American in space. They had trained together almost two years. The decision, Gilruth said, had been difficult because all of them were good pilots and had worked with great dedication. But the choice had been made. Shepard was to be prime pilot with John Glenn, as backup. Gus Grissom would fly the second Redstone, with Glenn again the backup.

Gilruth wanted the selection to be kept in strictest, confidence, trying to keep the pressure off any one individual. The press would not be informed until the day of the launch. Shepard told his wife Louise; some members of the family knew before he flew.

The flight was targeted for March 1961, but one more test flight remained. An unmanned Mercury capsule had already been lofted on a ballistic trajectory by a Redstone rocket. Now it was a chimpanzee's turn to pave the way for man.

His name was Ham and he had a rough flight; because of an electrical problem, the escape rocket fired in error and his capsule overshot his target area by 112 miles. The flight was referred to as the "Great Chimp Adventure." The engineers said they understood the problem and recommended that NASA proceed. *"I was all for it,"* Shepard said, *"but others took a more conservative approach and thought another unmanned flight was necessary. That decision pushed my launch date back to early May and probably cost me becoming the first man in space.*

On April 12, the Russians stunned the world by launching cosmonaut Yuri Gagarin into a once-around-the world orbit. *"We all were surprised,"* Shepard commented. *"We had no idea they were so close, and his one-hour, 48-minute trip was certainly a more ambitious undertaking than the 15-minute, up-and-down flight I was training for. Beating the Russians was always in the back of everyone's mind. Without that extra test flight we could have been first. I was very disappointed."*

Shepard's day finally came – May 5, 1961. Bad weather wiped out an attempted launch a few days earlier, even before he got in the capsule. After announcing the postponement, NASA disclosed that Shepard was the one selected for the flight.

NASA Administrator Jim Webb prepared three possible statements for the press: One if the flight succeeded, a second if Shepard had to eject but survived, and a third if Shepard was killed. I asked Shepard what he thought about these pre-prepared press releases and he said, *"As a pilot you are interested in number one: Does the thing have enough thrust to get off the ground? Number two: Is it going to accelerate beyond the point where I'm not going to be able to cope with it during launch phase? Having satisfied yourself that you're going to be able to do that, then really it becomes more a matter of being concerned about controlling the vehicle in case of a malfunction rather than the enormity. If a Redstone blows up you're still going to lose your life if you don't react properly."*

I asked Shepard if Louise and family were at the Cape for his launch. Shepard explained they had agreed she would not come to the Cape but stay in Virginia Beach.

In the May 21, 1961, *LIFE* magazine exclusive article written by Louise, she recorded:

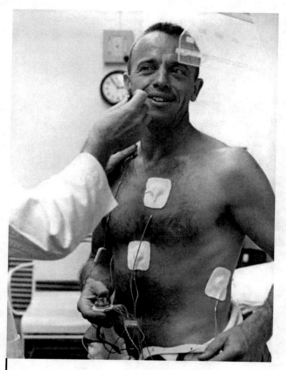

Shepard prepares for his ride

I last saw Alan almost two weeks before the shot. He had been busy for months working with the Mercury capsule that finally carried him so beautifully and had been home very little. So I went down to Cape Canaveral to visit him. Even there, staying in a motel at Cocoa Beach, I didn't see much of him, but at least we were able to have breakfast and dinner together every day. Through his natural ease and gaiety, I could feel his growing preoccupation with the flight that was ahead of him.

On my last evening in Florida, he drove me to the airport. We knew that we would not be seeing each other again until the shot was over and the conversation was a little strained. He was on guard against any display of emotion and was trying hard to be very casual. Not many words were necessary.

I understood how he felt. He was convinced that it would be better for the family and me if I waited out the flight in Virginia Beach. He would keep me closely posted by phone and I would be away from the pressures of the great build-up at the Cape. At first I had wanted to be near him when the shot was fired, but I decided to play it his way. On the surface our goodbye at the airport in Orlando was like a normal family parting. I didn't let emotion come into my thinking until I got on the plane. Then I had to let down a little.

Shepard continued to describe his pre-launch morning. *"Dr. Bill Douglas, the astronauts' physician, awakened me a little after 1:00 a.m. I shaved, showered and had a breakfast of steak and eggs with Douglas and John Glenn. John left for the launch pad to make sure my capsule was ready. I felt a few butterflies as Douglas gave me a brief medical exam and technicians helped me into my space suit. About 4:00 a.m. we left the crew quarters in Hangar S. Douglas and Gus Grissom were with me. I was pleased to see the skies were clear."*

I asked him, *"What was it like that morning as you climbed out of the transfer van? You seemed to be hurrying to catch your ride."* Shepard hesitated a moment and said, *"I was struck by the beauty of the Redstone topped by the Mercury capsule on the launch pad. I'd named the capsule Freedom 7, because I've always been a patriot and believed freedom was important to our country. The seven represented the seven Mercury astronauts."*

I asked him why he deliberately stopped and took a long look at that rocket bathed in searchlights with puffing white clouds of venting liquid oxygen. He explained he was never going to see this rocket again, so he stopped and stared up at it for a few seconds and then hurried to get aboard.

At 5:18 a.m., with the help of Glenn and some technicians, Shepard squeezed into the capsule's very tight quarters. Shepard describes the scene.

"John was my backup. One of his responsibilities was to be sure the spacecraft was exactly in the right configuration, all the switches in the right place, all the straps in the right place and everything. I came up and John was there in the white room. They literally stuffed me in, almost a shoehorn type treatment. There you are, locked and strapped in and almost totally immobile. I looked up on the instrument panel and John had left a note that said, 'There will be no handball playing in this area.' That took the tension off, I tell you."

Von Braun monitors Shepard's countdown in Cape Canaveral blockhouse with Capsule Communicator (CapCom) Gordon Cooper

The hatch was closed, and he was alone, shut off from the world except for radio chatter and a wide-angle periscope that gave him a distorted view of the outside. He remembered the butterflies were pretty strong. To counter the nervousness, he plunged into preparations, running through checklists, testing the radio systems and switches. I asked, *"Why were you nervous?"* Shepard looked at me with a cocky grin and said, *"I didn't want to screw the pooch."*

He was scheduled to lift-off at 7:00 a.m., but the countdown was repeatedly delayed—by an overheated power inverter, clouds that moved over the area and a high-pressure reading from the Redstone fuel tank.

Shepard had been in the capsule for three long hours. This was before NASA had designed and perfected urine collection bags or UCBs in astronaut jargon. It seems that Shepard had drunk a considerable amount of government issue coffee that morning before zipping into the silver Mercury spacesuit. They also had added electrodes to his body to collect data so he was really wired into the spacecraft, which in turn was wired into launch control.

As he reclined on the couch, he mentioned several times that he needed to relieve himself. The launch control personnel refused due to possible short circuiting in the spacesuit since it contained the electrodes measuring heartbeat and body temperature. After more pleading from Shepard, he was granted permission to, "wet his pants." Since he was lying on his back, the urine released soaked his rear end and halfway up his back inside his space suit. Nothing shorted out, the launch count continued and Shepard was a very happy astronaut. Actually this wasn't the first time a Mercury astronaut had wetted his spacesuit. All of them had spent hours in a centrifuge in Johnsville, Pennsylvania. and several had "let it go" in the suit rather than stop the test.

Shepard awaits liftoff aboard Freedom 7

I asked Shepard about his now famous quote, *"Light this candle."* Shepard explained, *"As the delays continued, I became a bit irritated. At one point I barked over the intercom, 'All right, I'm cooler than you are. Why don't you fix your little problem and light this candle?'"*

Finally, at 9:34 a.m., the candle was lit. As the Redstone engine flashed to life, Deke Slayton, who was capsule communicator (CapCom) in launch control, radioed, *"Lift-off. You're on your way, José."* And that's how Shepard got his call sign, "José," after comedian Bill Dana's character, the cowardly astronaut José Jimenez.

Shepard says, *"It was a strange, exciting sensation—a smooth, gentle rise off the pad. There was a lot less vibration and noise than I'd expected. About a minute after lift-off, the ride got rough as the rocket and capsule passed from sonic to supersonic speed, then sliced through a zone of maximum dynamic pressure as peak speed and air density combined. The acceleration kept building, pressing me into the seat. My vision blurred for an instant, but I was able to see the instruments and radioed 'all systems are go' to the control center.*

"At two minutes, 22 seconds, at my top speed of 5,036 miles per hour, the engines shut off on schedule," Shepard recalled. *"I heard a noise as a small rocket fired to separate the capsule from the booster. At this point my pulse rate, which was 90 before launch and 126 at lift-off, shot up to 138.*

After dropping the booster, Freedom 7 and I were weightless, and I felt the capsule begin its automatic 180-degree turn to get into position for the rest of the flight. The weightless feeling was pleasant and relaxing. It was a relief not to feel the weight of my body pressing against the couch. Washers and dust particles drifted out of crevices in the cabin. Although I was traveling at more than 5,000 mph, I had no sensation of speed because there was nothing to judge speed by. Through the periscope, I could see the sky, a very deep blue, almost black, because of an absence of light-reflecting particles."

Asked if he planned in advance to make a statement, Shepard commented, "Most of the time you're either launching or reentering with a lot of other things to do. The flight only gave me about 30 seconds in which to look around, peek through the periscope—the only way I could see the Earth— and make some profound statement. Fascinated by what I saw, I radioed, 'What a beautiful view!' I could see the coast of Florida, Cape Canaveral, Lake Okeechobee, some of the Bahamas, clouds over Cape Hatteras—really striking from more than 100 miles up.

"A little more than five minutes into the flight, the three retro-rockets fired at five-second intervals, briefly pressing me back into my couch. I didn't need the retro-rockets because I was on a ballistic course that would get me back to Earth, but we wanted to test them because they would be used on Mercury orbital flights to slow the capsules so they would drop out of orbit. I checked out the manual control system, maneuvering the capsule in roll, pitch and yaw. The controls were crisp and positive. Until then, the vehicle had been controlled by an automatic system. Thinking back on it now, if I had a laptop in that little Mercury capsule, I could have flown that sucker to Mars!" exclaimed Shepard.

Time passed quickly, as Shepard started to prepare to re-enter the atmosphere. "I aimed the bottom of the capsule down at about a 40-degree angle and switched controls back to automatic. At 230,000 feet, a green light came on, indicating that gravity forces were starting to build on the capsule. I braced myself because I knew the g-forces on re-entry would be much higher than the 6-g's I'd experienced on lift-off. Indeed they were, building to 11 times the normal gravity here on Earth. To assure the controllers, I radioed every 10 or 15 seconds that I was okay. When you get anything above 8-g's, you have to breathe in gasps so my voice transmission was pretty rough. But I was still saying, 'Okay,' and I could read everything all right."

The g-forces began to drop at 80,000 feet. At 30,000 feet the atmosphere slowed Freedom 7 to about 300 mph. "Deke (Slayton) told me I was right on target and that I should land in the Atlantic in the middle of the recovery area. At 21,000 feet, through the periscope I saw the drogue parachute come out; at 10,000 feet the main chute unfurled. What a welcome sight!" Shepard exclaimed.

Helicopter picks up Shepard with Freedom 7

"Freedom 7 hit the water with a good solid pop. But it did not seem any more severe than the jolt a pilot takes when his plane is catapulted off an aircraft carrier. The flight had lasted just 15 minutes, but in that time I had traveled 116.5 miles into space and had landed 302 miles southeast of the launch pad. The capsule flopped on its right side and water covered the porthole. It righted itself slowly, and I radioed I was okay. Within minutes a helicopter was overhead, attaching a shepherd's hook to the top of the capsule. 'Freedom 7, this is Rescue One,' the helicopter pilot radioed. 'You've got two minutes to come out.' I decided he knew what he was doing, and I opened the door and took a sitting position on the sill. The *chopper dropped a horse-collar sling, and I slipped it on, then was hoisted up and sank into a bucket seat as soon as I was onboard,"* he explained.

"On my seven minute ride to the aircraft carrier, I felt relieved and happy. As we approached the USS Lake Champlain, I could see the deck lined with sailors. They were all waving and cheering. I got all choked up over that. I guess it was because I was a Navy guy and had spent so much time on a carrier. That was really an emotional time," related Shepard.

"Boy what a ride!" Shepard told the captain when he was on deck. I asked Shepard why he climbed up and looked into the capsule after he got on deck. *"Well, for one thing,"* he said, *"a fighter pilot never leaves his helmet in the cockpit, so I reached in to get my helmet. I also took a look around the instrument panel to see if I turned everything off."* That was Shepard.

Shepard checking to see, "if I turned everything off."

I asked Shepard about Shorty Powers' commentary to the press at the Cape, repeating Shepard's famous comments at lift-off, *"Everything is A-OK."* According to Shepard those were not his words. Those were the words of the spin-doctor, quoted worldwide.

The Mercury system—rocket, capsule and man—had worked to perfection. The nation had demonstrated its manned space flight capabilities in the open, with the world watching. Shepard knew he had done a nearly perfect job. He had not screwed the pooch and the brotherhood was happy.

Alan Shepard's 15-minute sub-orbital spaceflight aboard Freedom 7 caused the visionary President John F. Kennedy to issue an extraordinary public challenge to the Russians in an historic statement made on May 25, 1961 during a joint session of Congress: *"This nation should commit itself to achieving the goal, before this decade is out, of landing a man on the moon, and returning him safely to the Earth. No single space project will be more important to mankind or more important for the long-range explorations of space; and none will be so difficult or expensive to accomplish."* Kennedy added, *"We take an additional risk by making it in full view of the world. As shown by the feat of Astronaut Shepard, this very risk enhances our stature when we are successful. This is not merely a race. Space is open to us now and our eagerness to share its meaning isn't governed by the efforts of others. We go into space because whatever mankind must undertake, free men must fully share."*

With Shepard becoming the first American to ride a rocket, the flight of Freedom 7 became an indisputable American success and the Kennedy administration had an American achievement. Shepard, his M7 buddies and wives, were invited to the White House for a Rose Garden ceremony with First Lady Jackie Kennedy and Vice-President Lyndon Johnson present. As President Kennedy started to pin the NASA Distinguished Service Medal on Shepard's lapel, he dropped it. Kennedy, with his quick wit, said, *"Alan Shepard, with the members of the House and Senate and the space Committee are with us today. This decoration has gone from the ground up."*

Later that day, the Shepards rode with Vice President Johnson in a motorcade down Pennsylvania Avenue. Thousands of people lined the street waving American flags. *"Johnson mentioned several times as Louise and I were riding with him, to get up on the back of the seat and wave at the crowd because, 'You are a famous man and that's how you behave when you are a hero,'"* Shepard remembered aloud. As they reached the steps of the Capitol, Johnson turned and said, *"Let me give you a little advice. Never pass up an opportunity for a free meal or a chance to go to the men's room."*

President Kennedy awards Shepard the NASA Distinguished Service Medal, stating, "This decoration has gone from the ground up," with the M7 looking on.

Did you anticipate that statement in President Kennedy's joint session of Congress address in May 1961 about landing man on the moon within the decade?

Von Braun leads celebration of Shepard's flight in Huntsville

Shepard: *No, actually an announcement was made about three weeks after the meeting that day. The NASA guys went back and came up with a proposed time schedule. It was about three weeks after the first meeting in Washington. Kennedy and the leading members of the Senate, the House and the top people in the NASA program were there. At that time, the president was so enthused about the flight and the response of the public towards the flight; he was sitting on the edge of his rocking chair saying, "What can we do NASA? Where do we go from here?" Now some people say it was a political decision. Well, obviously* anything coming out of the administration has to be considered historically a political decision. I sensed that Jack Kennedy was a real space cadet; not only from that moment, but also from the tremendous interest he had in ensuing subsequent flights. The NASA guys said they were making projections on pieces of paper back in a corner of the lab. They were talking about building bigger rockets and everything. Then Kennedy said, "Well, can we go to the moon?" The NASA guy said, "Yes, with the right amount of attention and energy we can go there." Kennedy replied, "Great, we're going to do it before the end of the decade." Well, the NASA boys got a little tight under the collar. They said they'd like to go to the moon, but we're not sure we could make it during that time frame. I didn't have an inkling, of course* (about the president's address). *We weren't privy to the discussions being carried on in Washington.*

Glenn: *In some ways you can look back in those days when we were making decisions to go from one program to another and liken it to where we stand today. We need a will to go ahead. In those days we knew enough about the engineering to design the new series of spacecraft. Today, we have the option to do the same thing with the permanently orbiting space station which expands our capability tremendously to do research. I'm afraid we lack some of the national will to do that. Back when John Kennedy set a goal for going to the moon, we didn't need a single scientific breakthrough. We knew that it was a mammoth engineering problem. There were lots of slide rules back in those days and the early computers to figure this whole thing out. It was engineering almost entirely. You didn't have to develop new concepts of rocket power, new metals and things like that, by and large. Today, it's the same thing. I think our next step should be to take everything we know and go to the permanently orbiting space station. I think we too often overlook the international geopolitical aspects of our space program. We've gone ahead of other nations around this world because of two very basic things. One is education; and the education wasn't just for the kids from the castles in this country, it was for everyone. We haven't done a perfect job with education, but we've done better than anyone else in this world. Secondly, we've put more back into basic research than any other nation. We've learned to do things first that form jobs, creating industry and employment. This led us to preeminence in this whole world. We've found the kids of the world are looking at our space program as an example of research and technology. They wanted to get their education and come to the United States of America and go to the Harvards, the MITs, Cal Tecs and all the different places in this country. We too often overlook* *that. I don't want to see the kids of the future aspiring to go to France or a university in Russia or some other place to get their great technical education. Let's move on with this thing and take the advances clear from Mercury up through Apollo, and retain that kind of leadership in the world. Waving your finger in the air and saying, "We're number one," after a ballgame is one thing. It's another thing to be number one and have the option of forming your own future as a nation. That's what this whole thing is about. We helped start this in the toddler stage. Now we have the option of moving out and setting up permanently orbiting space stations to do pharmaceutical and material research from space; onboard manufacturing. We can lead the world in these areas. Are we going to lay that down now and let it go to somebody else? We shouldn't.*

Shepard: *I'm not sure that we'll ever in our generation get the same amount of tremendous interest and support we had for the Apollo program. I think it's important for us to realize that space station may not be as exciting as a trip to the moon and back. It's incumbent upon us to tell the public, as John indicated, having a permanent space station is important and advances our technology and improves the value of life here on earth. Obviously we have to educate our young people so they grow up and get excited about this. Probably never quite as exciting in our generation as the Apollo program, but none-the-less, it is exciting and has to be conveyed to the public.*

When Kennedy issued the challenge, did you think, "I might go?"

Shepard: *Yes, I did. I'm sure all of us felt the same way. Deep down inside we're just ordinary people and there's a tingling excitement when you think about going to the moon.*

Shepard is the only one of the original group that got to the moon. How do the rest of you feel? Are you disappointed that you didn't make it to the moon?

Glenn: *Yes, I think we all feel that way. I was the oldest of my group. After my flight they were suggesting for whatever reasons for me to go into some areas of training and management and things like that. I chose not to and stayed around the program for awhile. I felt the upcoming flights should go to those who were going to be usable later on for some of those efforts. For me to stick around being the world's oldest permanent...used, second-hand astronaut might be wishful thinking. The way it's all worked out, perhaps I should've stayed around. I do regret not having been to the moon. I think that anytime one of these flights goes everyone in the line-up would like to be on the flight.*

Shepard: *John, I don't mean to take exception to what you said, but you said you **were** the oldest member of the group and you still **are**!*

We had our share of VIP visits at Huntsville's Marshall Space Flight Center in the 60s. The most prominent visitor was President John F. Kennedy, the man who took Wernher von Braun's advice to heart and committed the U.S. to an all-out race with the Russians to the moon.

President Kennedy decided he wanted to see the hardware and meet the people working on his moon landing program. For his first hands-on look, he picked von Braun's Rocket City, U.S.A.— Huntsville, Alabama. This was a major event in 1962 for Huntsville. It was big for Wernher von Braun too.

We received the call in the public affairs office that the president was coming to Huntsville. A full-duration, Saturn booster test firing was a must. My first job was to organize the people and hardware in the rocket hangar where the president would tour. Next, was working out the route he would take to watch the test firing including stops in-between. We wanted to take him to four locations, and the Army, the landlord on Redstone Arsenal, wanted three. We soon found out we were a little ahead of ourselves with the plan. A pre-visit by White House staff and Secret Service established who was in charge. We showed them our planned route and activities and soon learned they had other ideas. The Secret Service told us where we could take the president and the White House staff told us what we could do with him. Even though we were on a military installation, the Secret Service combed the place looking for hazards—places that could be used by a sniper, especially on top of buildings. We had to provide a list of the names, dates of birth and the Social Security numbers of all persons who would be in contact with the president's party.

President John F. Kennedy with von Braun during visit to Redstone Arsenal and NASA-Marshall Space Flight Center, Huntsville, AL. in 1962

After several more visits and meetings, a plan was approved for Kennedy's walk through the rocket hangar and the test firing demonstration. The Army was approved for one stop at their Guidance and Control Lab.

For us, the test firing approval was big. At first the Secret Service said, *"No way,"* because of safety concerns. But after we convinced the staff that this would be **the** event of his visit, approval was granted for the president to observe from Heimburg Hill, 2,000 yards from the test stand. Hard hats were a big discussion point. The safety people were adamant and vocal. That's when I learned Jack Kennedy didn't like hats; a point made very clear by his staff. I don't believe Kennedy wore a hard hat the entire visit, although an aide stuck one in his hand just before the firing.

It was the president's first exposure to a NASA center. It was his program and he flew on Air Force One, accompanied by the vice president, the NASA administrator, the U.S. secretary of defense and all the ranking space people on the Hill. The White House press corps arrived in a separate plane. Nearly 200 press people—representing all major networks and news services in the country—were on hand.

Kennedy came to see the moon rocket and meet the man and team tasked with delivering the next frontier. It was a fascinating experience to observe Kennedy and von Braun—two men from such different worlds—as they became better acquainted. Rocket man von Braun did more than anyone else to convince President Kennedy and the leaders of Congress that this nation was capable of undertaking a complex and dangerous mission like landing a man on the moon and returning him safely to Earth. Dr. von Braun met the president and his party at the Redstone Airfield with Army officials and full military honors, including a 21-gun salute.

After a speech by President Kennedy to foreign military students, the party traveled to the Army's Guidance and Control Lab for a short tour and demonstration of various classified missile systems. From the time they climbed in the open top limo, to the end of the ride throughout Redstone Arsenal, Kennedy and von Braun hit it off.

Next, they proceeded to Marshall's rocket hangar where the 152 foot-long Saturn I was displayed along with an F-1 engine that would power the Saturn booster. Several hundred employees were present and when the president and his party stepped on the floor of this giant rocket hanger there was thunderous applause. Then it became very quiet as von Braun began his enthusiastic explanation of the rocket hardware. You could see the chemistry working, each respecting the other's abilities and understanding the vision they shared. It was important to Kennedy to personally be reassured by von Braun that we could meet the commitment Kennedy had made to the people of this country. He asked that very question of von Braun and received a firm answer, *"Yes, Mr. President, we are going to meet your commitment of landing a man on the moon and we're going to do it within the time you set."*

Last stop on the tour was the test division where a Saturn I booster was being prepared for a test firing. We planned a full duration test—two and a half minutes—running the eight engines at full throttle. Marshall was the place to come if you wanted to see the latest happenings in the Saturn program. We were not only building the Saturn boosters, but ground testing them and this was well before anything was happening at the Cape or Houston. We couldn't launch a Saturn moon rocket, but we could put on the ultimate fire and smoke show.

This was when von Braun was at his best, taking high government officials, congressman, senators and presidents through the Huntsville rocket factory. He would take them through an animated flight to the moon—make them feel as if they were going—gesturing, pointing and describing how the big booster worked. This would be followed by a visit to the test stands where the boosters were ground tested. This is where Kennedy was, on Heimburg Hill, watching a Saturn booster ignite, shooting smoke and flames skyward, generating 32 million horsepower.

After the firing, Kennedy was excited. He grabbed von Braun's hand, shook it, congratulated him and beamed with that Kennedy smile. As the president described the firing he said he could feel the ground and structures shaking nearby. He could feel the vibration in his chest, the shockwave and heat as it hit his face through the slit in the front of the bunker. He pointed down to his feet and said he felt heat coming up his pant's legs. The president of the United States had just witnessed a Saturn firing, one of the most powerful machines man had ever built, and he was impressed.

Kennedy and von Braun climbed in the limo and drove into the restricted test area to have a closer look. As they drove by the test crew at the blockhouse the president told the driver to stop. He asked von Braun, *"Who was in charge of the test I just saw?"* Von Braun asked the test crew. Word passed up through the crew, *"The president is asking about you, Bob."* Bob Saidla, a young test conductor, came running toward the motorcade and the crowd practically propelled him into the limo, causing the Secret Service some concern. He approached von Braun's side of the limo. The president, who is now standing in the limo, reached across von Braun and grasped Saidla's hand and congratulated him and his crew on a most impressive firing.

Kennedy was convinced, more then ever, that his goal of landing an American astronaut on the moon first, beating the Russians, was going to happen. He repeated his pledge that the U.S. would explore space. *"I know there are lots of people now who say, 'Why go any further in space?' When Columbus was halfway through his voyage, the same people said, 'Why go on any further? What can he possibly find? What good will it be?' I believe the United States of America is committed in this decade to be first in space, and the only way we are going to be first in space is to work as hard as we can, here and all across the country..."*

Hundreds of Marshall employees lined the road as the motorcade returned to the airfield for a midday departure. He was so impressed with the enthusiastic team and the work he saw going on at Marshall and the Saturn booster test that day, he invited von Braun to accompany him on Air Force One to visit other NASA centers. Von Braun said he continued asking questions about the Saturn and how the trip to the moon would to be accomplished. They got to know each other very well. We sent von Braun's briefcase and toothbrush on another plane. Before boarding Air Force One, Kennedy took the time to shake hands and thanked each of us who had worked on his visit to the Rocket City.

Shortly after that visit, President Kennedy and First Lady Jackie Kennedy extended an invitation to Dr. and Mrs. Von Braun to attend dinner at the White House. Unfortunately, the date of that dinner was one week after President John F. Kennedy was assassinated. The president who challenged the United States to land man on the moon within the decade never had the opportunity to see the Saturn moon rocket fly, or one of his astronauts walk on the moon.

The "astronaut in a barrel" was a great concept—unless you happened to be part of the brotherhood. Shepard, who had encountered an inner ear problem and was no longer on flight status, was now the leader. As the chief of the astronaut office, Shepard became a feared and dreaded obstacle to public relations people. He often told us, *"You are the last to know and the first to go."* So began the battle of the NASA public relations people with Shepard, the icy commander.

After the M7 astronauts were selected, a clamor arose for press interviews and appearances. Requests were being received by all NASA organizations for a live, walking, talking astronaut. Once Shepard and the other guys flew, the demand grew even greater. Getting the astronauts to participate in press and public events was like pulling teeth. In fact, it became nearly impossible. In most cases, the answer public affairs received upon making a request was, *"No, we're too busy training for our next flight."*

This didn't sit well with us in the public relations ranks. NASA was losing too many opportunities for good press. As noted earlier, the M7 had a contract with *Life* magazine for exclusive stories on their lives and flights. From the astronauts' standpoint, obligations had been met. Consequently, these additional press appearances and interviews were unnecessary. Obviously, this made other members of the press corps a bit unhappy because, "NASA's widest possible dissemination of information of the space program," should include astronauts. A solution was offered that each week one astronaut would be offered up, or as Shepard preferred to say, "sacrificed," for interviews, appearances and general public relations tasks. This helped because now those of us in public relations knew we had at least one astronaut each week.

To get the most out of the process, we proposed to higher-ups an astronaut training course on how to conduct yourself upon becoming an astronaut. To tell the truth, the astronauts weren't the only ones who needed the course; we did, too. We also proposed that NASA shouldn't use the term astronaut loosely: wait until the candidates do something before bestowing the title.

The Air Force must have heard us. Although the Air Force only had about 20 of their breed in the first 72 astronauts, they considered their organization the prime training ground for producing astronauts for NASA. Our Air Force friends created a "charm school" for its leading candidates. They were taught how to act, dress, stand, drink and talk in public. Astronaut Mike Collins, a former Air Force pilot, said, *"They told us we must wear socks that covered our legs up to the knee so when we were seated, we didn't show our hairy legs. They showed us how to stand with our hands on our hips with thumbs pointed to the rear like fighter jocks."*

Among the other rules cited: It was permissible to have one drink while attending receptions and cocktails parties, as long as it was a "man's drink"—bourbon or scotch. It certainly couldn't be wine or some fancy mixed drink. When carrying books or manuals, they had to be carried in one hand at the side, never held against the chest. While traveling in a vehicle, an astronaut should never be seen letting a female drive. Aviator sun glasses were acceptable with any dress code. Crew cut was the hairstyle and no facial hair allowed.

Convincing Shepard that an event was significant enough for one of his guys to actually make an appearance was the next problem. For those of us who had to deal with Shepard, the permission process was not much fun. When asked the question, *"Can I have an astronaut this week for such and such?"* the answer was, *"No, my guys don't do old lady club speeches,"* or *"We already gave that (expletive) an interview and he misspelled Wally's name and thought re-entry was some kind of sex thing."*

Public affairs people weren't the only ones finding it difficult to bond with the icy commander. Even members of the corps had their trying moments. Owen Garriott was one of the first scientist-astronauts named to the astronaut corps. He flew on Skylab and Spacelab. While returning from flight school, he found his parking place taken at the astronaut's office, Building 4. Since he wasn't going to be long visiting, he slipped into another marked space near-by. Upon returning to his car, he found Icy Commander Shepard's business card on his car with a handwritten note saying, "Please steal someone else's parking place the next time."

Some time later, Shepard invited Garriott to fly to New York with him in one of NASA's T-38's. On the leg Garriott was flying the aircraft, he began looking out of the cockpit to checkout the terrain and interesting landmarks. This required rocking the wings back and forth 10 degrees or so to see. As they continued on to their destination, Shepard said, *"What are you trying to do, Owen, make me sick?"* Garriott considered that a compliment, as if he could make the icy commander, naval aviator and test pilot a bit uncomfortable.

We had the Cold War, U2s, H-bombs, and the Cuban Missile Crisis: In that context, did you view the Space Race as a possible scientific or ideological contest?

Shepard: *I viewed it from the standpoint of the scientific benefit of expanding space technology, which we had already created. There is an interesting thought, however, having to do with the fact that Kennedy had announced the moon project so quickly after the first flight which I made in 1961. As a matter of fact it was only three weeks after that first 16-minute flight when he committed the nation to landing on the moon within the decade.*

Do you feel unique when you look at the moon in the sky?

Shepard: *I don't think I'll ever lose that sense of being there on the moon. It was so unique; certainly, one will always remember that kind of experience. I don't even have to look at the moon to remember it. Things that happen during daylight without looking up there remind me of funny things that happened up there or serious things. I'm sure those thoughts will be with me for as long as I'm around.*

When you're playing golf do you ever remember that swing you made on the moon?

Shepard: *Well the thing that intrigued me, being a golfer, was that the same speed of the club through the ball was going six times as far. Meaning, my tee shots could be approximately 1,500 yards. Then, the fact there is no atmosphere, so if I spin the ball it won't curve to the right or left, no slicing or hooking…then, the fact that the ball is in the air six times as long. That one-hand, chili dip, six iron shot I hit should have gone about 30-35 yards, but it went over 200 yards and stayed in the air for almost 30 seconds. It was a real slow motion deal. It was very exciting and the interesting thing was I didn't have time to collect the golf balls. The deal I made with the boss was this would be right at the end of the last excursion, outside the lunar module. I didn't have time to get the golf balls, but the club—a Spalding six iron—was collapsible, specially designed. Now it's in the golf museum (United States Golf Association Museum, Far Hills, New Jersey) for all the golfers in the world to see.*

I remember you said that you just might return to 1/6 gravity to help your golf game.

Shepard: *Listen, I have more trouble with my golf game on Earth than I did up on the moon.*

During the countdown of the Mercury flight, what did you think your chances were of completing a successful mission?

Shepard: *Reflecting back to that first Mercury Redstone flight, all of us realized there was a risk and missiles had blown up. We also sensed the reason for the missiles exploding, to some degree, was* maybe a lack of attention by the people who put them together; maybe a little bit of haste in the preparation. The fact that man was going to be on the top of that rocket, in that spacecraft, sort of brought the team together. Somehow they worked longer hours—above and beyond the call of duty—to be sure everything was exactly correct. That team would, in fact, create the very, very best atmosphere possible for manned flight. By that, I mean the chances of failure were reduced by NASA training us so hard. I think we sensed that and we went along with it and risked our lives.

Scale model of Shepard's surprise golf shot on the moon

Your mission took place only three weeks after Yuri Gagarin's orbit of the Earth. Did NASA rush putting America into space to save face?

Shepard: *As far as the Space Race is concerned, we were certainly aware of the fact that the Soviet cosmonauts were training as we were. The general substance of their program was known to us, but the specific schedules were not known. Whether or not it was known somewhere in the secret system, the black world, I don't know. The point is we were not designing our training and launch schedules to compete with the Soviets—in general. Decisions were made based upon the engineering values and the time required for a particular training phase. Of course, when Gagarin flew before I did... there was some disappointment, but the die had already been cast for my flight. The ground rules had been made. Those ground rules were not changed. The program was not really accelerated as a function of Gagarin's flight. It was the correct decision to make. The engineering decisions really should say we'd go when we're ready. Beyond that I think Mercury was really proven quickly and effectively. Our objectives in Mercury were very simple: to put a human being in space and orbit; to assess his ability to control and respond and be rational most of the time. Those goals were reached surprisingly quickly. We had only six Mercury flights, two sub-orbital and four orbital. At the end of that we said we'd reached our goals and we did it in a very short period of time, approximately two years. So the contributions were substantial. All doubts about man being able to perform in weightlessness and zero gravity were dispelled and we moved rapidly to the Gemini program.*

Shepard shows off his famous six iron shot on the moon crater at Space & Rocket Center, 1989

Something went wrong on almost every Mercury flight. How do you cope with knowing America's space industry has worked terribly hard to make sure everything is foolproof? In your case, did you have any doubts?

Shepard: *To me, the true value of man in space was demonstrated in Mercury and has been demonstrated subsequently very many times. It's a fact of life that with hundreds of thousands of parts, something is going to go wrong. Hopefully, what goes wrong isn't catastrophic: A switch that has to be replaced or switched off by the astronaut. Certainly, in Mercury we found that the value of putting a man in space was not only his judgment and observation, but the fact that he was a back-up for so many of these systems. We've seen in the Soviet program where they still plan things much more automatically than we do. They don't really take the true value of the cosmonaut in the loop. I think we've seen many more failures in the Soviet program because of that attitude.*

After becoming the first American in space, you finally realized your original dream, the only Mercury astronaut to walk on the moon. What were your thoughts about the early days when you finally set foot on the lunar surface?

Shepard: *My expectations, my motivations and my objectives in the space business were really those of everyone. That is as long as you qualify to fly and you're physically ready to fly, you want the very next flight available. For a number of years I was physically disqualified and grounded, but got that corrected. Fortunately, I was assigned to the very next flight available, the lunar mission. To me that was an accomplishment in not only the sense of going there, but personally being able to overcome some physical difficulty and make the mission to the moon. From a personal standpoint, there was a tremendous amount of self-satisfaction involved. Obviously from a historical standpoint, being allowed to do that and being a part of that program will always be important in my life. Because certainly, for two or three generations anyway, the moon landings will be considered one of the most exciting things that ever happened to man.*

"Levity is the lubricant of a crisis. We resort to jokes, pranks and good natured kidding to relieve tension, stress and boredom," said Schirra. "Gotcha," meaning "I got you," was made popular during the beginning of the Mercury program. Shepard and Schirra were the instigators of many jokes—gotchas—some of which were downright nasty and others funny.

It's important to note, however, it didn't take a crisis to generate a good gotcha by any of the M7 astronauts, especially Schirra! A case in point, Schirra and crew hosted a dinner to celebrate Shepard's fifth anniversary of his Mercury Freedom 7 flight. Bob Gilruth, Max Faget, Chris Kraft, Wernher von Braun—all the big names in manned spaceflight—and the M7, of course, attended. Schirra was the emcee and introduced a special production, "Alan B. Shepard, Jr., Hero or How to Succeed in Business Without Really Flying Very Much."

He portrayed Shepard as the space hero—with only 15 minutes of space flight—walking around and looking into his spacecraft after recovery of Freedom 7. Old films of flying contraptions failing to get off the ground or falling out of the sky followed, indicating Alan B. Shepard had not been very successful in getting himself and his machines off the ground. Only two had gone before Shepard: chimps Ham and Enos. As the film neared conclusion, famous American heroes including George Washington, Albert Einstein and Abraham Lincoln appeared on the screen and the voiceover intoned, *"Every great man is led by another great man. Behind this great man, Alan Shepard, was none other than Captain Wally Schirra."* Schirra, decked out in a white wig, took center stage. The room erupted in guffaws and snickering, but Shepard did the prank one better. He found out about Schirra's gotcha and a voice announced, *"Wait a minute, Wally. There's more."* Shepard appeared on screen, also sporting a white wig, and said, *"I will see you in my office tomorrow at 0800, Schirra, to discuss the rank of permanent captain, and that's an order."* As Shepard was introduced, the microphone went dead and everyone exited the room, leaving Shepard alone at the podium. Another Schirra gotcha: It became an all-time great.

The Redstone was often compared in size to monuments that were similar in size such as this one at Columbus Circle in New York City. Here the monument is launched and the Redstone stays on the ground.

The "Lighthouse That Never Fails" caper was popular in the early days of missile and rocket launches. Many first time film and TV crews to the Cape had no idea which tower or structure to point their cameras to record these unique news-making events. To initiate these newcomers to this special fraternity of rocket watchers, those of us at the press site would often suggest that they set up to cover the blastoff—a term used in the early rocket business—in the direction of the lighthouse. It became such a popular gotcha at the press site someone made a film showing an Air Force sergeant climbing to the top of the lighthouse as it was launched like a rocket. The tower blasts off with the sergeant holding on for dear life. The Cape was shown disappearing beneath the ascending rocket. At the conclusion of the film, a gentleman was shown at a launch control console. When asked if he was an experienced rocket launcher, he commented, *"No, that was my first shot. I'm really a lighthouse keeper but I like to think for myself."*

It became quite common during Mercury's early days to find people gathering around the guys any time they showed up at a local watering hole. It wasn't hard to the find them. They were the ones wearing Ban-Lon shirts and aviator glasses attracting camp followers who wanted to be M7's extended family. A lot of people wanted to hang out with astronauts. Schirra, always the creative one, decided to form an unofficial connection between the brotherhood and the growing multitude of supporters. The Turtle Club was dreamed up by test pilots during WWII. Membership was diligently sought after and highly esteemed by those lucky enough to be initiated. Adherence to the creed and always giving the password when asked are the only responsibilities placed on the membership. Life was much more fun and took on a new meaning if you were a 'Turtle." The Interstellar Association of Turtles believes that you never get anywhere in life without sticking your neck out. Every turtle member presumably owns a jackass, and when asked, "Are you a Turtle?", you must answer with the proper password, or buy a round of drinks of his choice. Schirra and Shepard formed the Outershell Division of the Turtle Club, an exclusive organization that became quite popular in the 60s while the Apollo team was trying to maintain an even strain and beat the Russians to the moon. Limited to adults, members were indeed an illustrious group and included in its ranks some of the country's foremost leaders in the fields of government, finance, entertainment, aerospace, print and broadcast media and all others where aggressiveness, a feeling for fair play, clean thoughts and a sense of humor are keys to success. My membership card reads:

Interstellar Association
of
TURTLES OUTERSHELL DIVISION
This is to certify
Ed Buckbee
is a member in good standing and will remain so as long as he continues to give the password when asked by a fellow turtle.

Signed by Alan Shepard, High Potentate and Wally Schirra, Low Potentate

As a member in good standing you can subscribe new Turtles as follows:
We assume all prospective Turtles own a jackass. This assumption is the reason for the password. This password must be given if you are ever asked by a fellow member, "Are you a turtle?" You must then reply, "You bet your sweet ass I am." If you do not give the password in full because of embarrassment or some other reason, you forfeit a beverage of their choice. So always remember the password.
To become an official Turtle you must first solve the following riddles:

1. What is it a man can do standing up, a woman sitting down and a dog on three legs? (Answer: Shake hands)
2. What is it that a cow has four of and a woman has only two of? (Answer: Legs)
3. What is a four-letter word ending in K that means the same as intercourse? (Answer: Talk)
4. What is it on a man that is round, hard and sticks so far out of his pajamas you can hang a hat on it? (Answer: His head)

You are now a member of the Turtle Club. Govern yourself accordingly and procure new members.

A favorite pastime of Turtle Club members was to catch another member of the club in front of an audience and pull the old "check your six, check your fly caper." Schirra and I caught Shepard speaking at a corporate stockholders meeting in Huntsville. We took seats directly in front of him and began pointing and staring at the Admiral's fly. Shepard had no podium to hide behind so he performed some interesting movements on stage, trying to check and/or cover his fly. He became so distracted that he forgot what he was talking about. Chalk one up to the Turtle Club membership.

Wally Schirra scored one on the medical cult while living in Hanger S at the Cape. Wally had an extended session in a spacesuit, helmet and all, going through a complete spacesuit test. It was a long day. The medical people couldn't wait to get his specimen after this long stressful day. When he finally got out, he found a specimen beaker and filled it with Schlitz beer. The medical cult was in a panic for days.

During Schirra's first space flight aboard Mercury Sigma 7, CapCom, the guy in mission control who is the capsule communicator, Deke Slayton shot a terse question to Wally during the Atlas boost phase of his flight, *"Hey, Wally, are you a turtle?"* This was "live" and being recorded as "official commentary." Wally's response was, *"Rog."* Deke and the rest of the M7 thought he had gotten Wally for a free round of drinks by not hearing the correct answer. After recovery with all the crew gathered, they couldn't wait to harass Wally. Wally flips on the private flight recorder and there it was: *"Wally are you a turtle...You bet your sweet ass I am."*

When Deke made his flight aboard the Apollo Soyuz with the Russians, Wally had the opportunity to ask the same question of Deke and there was no answer over the "live" radio commentary. After Deke's return, the brotherhood gathered to congratulate and dog him about his silence when asked the big question. Deke activated his flight recorder and played it: *"Are you a turtle?"* Then Deke's answer, *"You bet your sweet ass I am."* And another great gotcha was recorded in the annals of manned spaceflight.

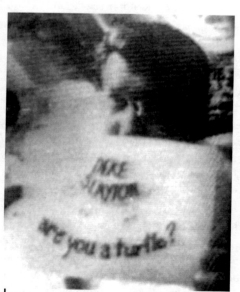

Schirra asks Deke Slayton, "Are you a Turtle," by flash card from Apollo 7. You bet he gets his answer

Schirra is most proud of the one he pulled at the White House. After Gordo Cooper's flight, M7 was visiting President Kennedy. This time it was the president's turn and he nailed Schirra. *"By the way, Wally, are you a turtle?"* Schirra thought twice before saying, *"You bet your sweet ass,"* to the President of the United States.

When Schirra teamed up with Cocoa Beach Hotel czar Henri Landwirth, they produced some really great gotchas. The Mongoose Caper was talked about for years in NASA public relations circles. The drill included Schirra and Landwirth selecting a time when a lot of visiting press and aerospace contractors were at Landwirth's Holiday Inn at Cocoa Beach for a launch. Schirra would walk through the bar holding a bloody towel wrapped around his arm saying, *"I caught him but he nearly tore my arm off."* There would be immediate interest as the bar-goers grabbed their drinks and followed the pranksters to Schirra's room to see what Landwirth had proclaimed to be "the biggest mongoose he had ever seen in Florida."

As the story goes, one giant-sized Yankee reporter was selected to be the first to view the varmint. As the crew crowded into Schirra's room to observe, Schirra pointed to a blanket-covered object on the bed and said, *"I think I knocked him out. Be real careful."* The guy pulled the corner of the blanket back and BAM! A spring-loaded hairy varmint with teeth flew out of the box. The victim and fellow bar-goers fled the room as Schirra had a laughter attack that brought tears to his eyes. Next step, he stuffed the mongoose back in to the special "jack in the box," readying it for the next victim.

Shepard, Cooper and Grissom visited their Corvette supplier and racing buddy, Jim Rathmann who liked gotchas too. Rathmann wanted to show his brave astronaut friends his rare mongoose, found only in Florida. Cooper had a snake with him in a bag that he used to scare Rathmann who ran out the back door. When he returned he said it was his turn to show off his mongoose. Rathmann, Grissom, Shepard and Cooper all gathered around the mongoose cage. Cooper, who's holding the snake and bag in one hand, reaches in the cage and the mongoose is launched like a jack-in-the-box. Cooper throws the snake and it lands around Gus' neck. The fur-covered mongoose lands on the floor near Shepard who tries to kill it with a ruler. So much for the brave members of the brotherhood.

Shepard was popular on the TV and radio talk show circuit. Sometimes we booked him in too many places in which case I would offer to substitute. I was standing in for him on a TV talk show as director of Space Camp at a remote station somewhere in the hinterlands at O dark 30. That is very early. Suddenly, half way through the interview, the host succumbed to an ill-timed fit of coughing. As the host staggered offstage, gasping for breath, the cameraman zoomed in on me, motioning frantically for me to continue the interview alone. In the best can-do tradition, I finished my presentation about Space Camp with an occasional glance in the direction of the now-spasmodic host.

As the subsequent station break neared its end, the host showed no signs of recovering. Clearly, this was one of those times that demand decisive action. The cameraman looked at me and said, *"Nice job on closing out your interview. You know we're the only ones in the station right now and I've got to operate the camera. Would you mind staying on and interviewing our last guest?"* I said, *"You must be joking."* He wasn't.

With some misgiving, I reluctantly agreed to interview the lady. How difficult could it be? Just ask her a question or two. She'll do all talking, the cameraman suggested. After stumbling through a few awkward preliminaries, I turned to the smiling lady sitting opposite me and improvised, *"Why don't you tell the viewers out there what it is your organization stands for."*

To my consternation, the lady snatched up an infant well hidden from me in a crib just out of camera range and proceeded without hesitation to demonstrate the proper procedure for breastfeeding. The lady was on TV representing, "The Society for the Preservation of Breastfeeding Babies." I remember it was a long 15 minutes. I managed to get through it somehow but it turned out to be my one and only time of acting as a TV talk show host. I told Shepard about it and he volunteered to do some public service announcements on behalf of the society.

One of the favorite props of TV talks shows was the astronaut suit. There was something about astronaut spacesuits that attracted TV personalities. On one show I dressed the co-host in a spacesuit complete with fish bowl helmet. I said to him, *"Whatever you do, don't even think about putting the helmet on because you will seal yourself in the suit with no way to breathe. Just carry the helmet under your arm."*

The host opened the show introducing me, followed by an introduction of the co-host dressed in the spacesuit who comes on the set and takes a seat next to me. We talk about the spacesuit, Space Camp and show a video. As I narrate the video, I notice out of the corner of my eye, the co-host has put on the helmet. Problem, yes, he has no air to breathe because he has sealed himself in the suit. As I continue trying to be cool and talk about Space Camp, I can see the helmet is beginning to fog-up. The guy is getting no air. He can't breathe. The host is still talking to me while the co-host is suffocating in my astronaut spacesuit on live TV.

As the guy tries to remove the helmet, it becomes stuck and won't come off. Now the host sees what's happening and begins talking to the guy who not only can't breath, he can't hear. He mumbles something that sounds like he's speaking to us from deep in a well. I try removing the helmet the prescribed way—unlatching the helmet ring—but it's jammed and won't release. I finally hit it with my fist to knock it loose and off his head. He gasped for air, and says, *"Now I know what it's like to be lost in space."*

I survived two appearances on the Late Night Show with David Letterman. I was on the show with Space Camp training equipment and some helpers dressed in flight suits. David enjoyed referring to us as: "Ed Buckbee and the Buckaroos." For some reason, he loved my name, repeating it several times. There were some challenges and what could have become career-changing experiences. While in the Green Room—where guests wait before going on the show—I became nervous and had to go the bathroom. As I stood in the men's room contemplating becoming a celebrity in the next few minutes, the door bursts open and the cameraman rolls in a camera. I'm standing there thinking that this is it! It's all over, career finished. I'm featured on national TV in the men's restroom. *Relax. We were just playing around. This was David's gotcha,* the cameraman laughs.

On national TV talk shows selling Space Camp

Buckbee with Monkeynaut Baker on Dinah Shore Show

Buckbee with James Brolin, Gordon Cooper on Mike Douglas Show

Buckbee with George Carlin on the Mike Douglas Show

Buckbee with Regis and Kathy Lee show

Buckbee with David Letterman, The Late Night Show. Letterman noted that Buckbee, who had asked for two signed photographs, was only entitled to one picture

Buckbee with Good Morning America producer and David Hartman

Buckbee welcomes Nichelle Nichols of Star Trek fame to the U.S. Space & Rocket Center

First Lady, Rosalynn Smith Carter trying on honorary astronaut hat presented by Buckbee

Buckbee and Senator John Kerry with full size Space Shuttle

Virgil I. "Gus" Grissom was shy around people he didn't know. He was the smallest of the group at 5 feet, 7 inches. Naturally tough and determined, he never thought of himself as a hero. He was born April 3, 1926, in Mitchell, Indiana. During WWII he served in the U.S. Air Corps as an aviator but didn't fly combat. After the war, he left the military and attended Purdue University where he earned a mechanical engineering degree.

Grissom rejoined the U.S. Air Force during the Korean War, and was highly decorated having completed 100 combat missions. He became an M7 astronaut in 1959 and brought Air Force piloting skills to the brotherhood. He was considered by some as "rough around the edges, but a good stick and rudder man."

He flew his first mission in July 1961 aboard a Mercury Redstone, the second sub-orbital mission. His capsule, Liberty Bell 7, was lost when the hatch prematurely blew causing the capsule to sink.

Grissom shown with his spacecraft, Liberty Bell 7

"I was just lying there, minding my own business when, POW, the hatch went. And I looked up and saw nothing but blue sky and water starting to come in over the sill. It was especially hard for me, as a professional pilot. In all my years of flying, including combat in Korea, this was the first time that my aircraft and I had not come back together. In my entire career as a pilot, Liberty Bell was the first thing I had ever lost," he commented with a sigh.

It was lost for 38 years—until a salvage crew recovered the craft and restored it for public viewing. Grissom's Mercury capsule was recovered in 1999; however, the hatch was never found. Grissom was awarded the Congressional Space Medal of Honor posthumously.

Liberty Bell 7 after recovery in 1999 on tour at
the U.S. Space & Rocket Center

What do you remember about your fellow M7 astronaut Gus Grissom who lost his life in the Apollo 1 fire?

Schirra: *I'd like to pick up on something on Gus and the travesty that the movie* The Right Stuff *showed him as a babbling coward in a spacecraft. When I came aboard the aircraft carrier on my Mercury flight it was some time after Gus had allegedly blown the hatch. I'm quite convinced he did not. Well, I asked to be hoisted aboard, not with just that idea in mind, but I didn't want to get my head bumped on a helicopter swing. So, we hoisted the spacecraft up on what we called number three elevator. While on that elevator I elected to blow the hatch rather than have all these bolts to remove the hatch…They had a mattress outside so it wouldn't fly off the ship somewhere. When I hit that button the recoil cut through my glove into my hand and made a big cut. The fellows* (M7) *arrived on the carrier after that with a big welcome aboard for my orbital flight. I held up my hand and said, "Gus, look at the cut I have on my hand. You didn't have anything like that at all on you did you?" The biggest smile I ever saw on the Pacific Ocean was on Gus' face which vindicated the fact that if he had hit that button inadvertently, he would've had a real welt on whatever part of his body hit it. I think that should be quick to put to rest, once and for all, that ridiculous story. Gus did not do that and a movie that would portray Gus as that kind of man should be banned. Gus helped me, to no end, backing me up in Gemini as I backed him up in Gemini. We worked very closely on the Apollo flight. I was his next door neighbor and then I had to be the executor of his will. He was a very close friend to all of us.*

Slayton: *The three of us from the Air Force knew one another pretty well. I had been friends with Gus for a number of years before the space program started. We were in a couple of schools together so I knew him quite well. He was a very capable guy and certainly different from the movie Wally mentioned. They seemed to be wrong in several areas, but particularly in regard to Gus. We don't feel* *that he lost his spacecraft or that he goofed. We think there is pretty firm proof that he did not blow the hatch. Certainly, he would not have been given the first Gemini spacecraft and Apollo spacecraft if he really had goofed. We miss him.*

Carpenter: *Gus didn't say a lot, but when he did speak, it was worth listening to.*

Shepard: *I remember being impressed with Gus as an individual and as a pilot. We Navy guys didn't think the Air Force guys knew how to fly and vice-versa.*

Glenn: *Marines don't think either one of them know how to fly!*

Shepard: *Marines are in a class by themselves and we're not sure where that is, yet! Gus was willing to join the group and he pulled his weight in the boat. He didn't talk too much, but when he did, as Scotty said, he really meant what was said. He was considered a great pilot, enthusiastic and competitive like the rest of us. I guess the thing I'll remember the most about Gus was the sacrifice he made for the program. Now, I think in the early days, including Gus, if we had said in the next 30 years we're only going to have two accidents, we each probably would have said I hope we're that lucky recognizing that human frailties were involved. I think what we saw in the tragic accident was NASA— a very successful organization having completed a great Mercury program, a great Gemini program and getting ready for Apollo—perhaps became a little complacent and smug. Perhaps some of the decisions were made hastily. The human element and frailties come to the surface and we have a tragic combination of things, which took the life of our buddy. But in the final analysis, that sacrifice reawakened the motivation and redirected the dedication. As a result: a marvelous, tremendous Apollo program.*

Slayton: *I think Gus did two major things in the program that he's never really received credit for. First, in the Gemini program he probably, single-handedly, was the program manager for M7 and made that a manned machine. We started in Mercury with a machine designed for chimpanzees because they didn't trust man in it. We had to convert it a little bit to get a man backup. We put Gus into the Gemini program—as the crew interface manager—very early. We nicknamed Gemini the "Gusmobile" because he was so much a part of that spacecraft. We ended up with a machine that was really a manned machine in Gemini, one that could be piloted. Most of the major unknowns relative to getting into the Apollo program came out of the Gemini program. Rendezvous, for example, extra-vehicular activity, long duration flights and guided entry; all of those things were worked out in the Gemini program. So that was the first thing. The next thing Al talked about—the fire. There were an awful lot of things wrong with Apollo and we knew it at the time of the fire. We made a conscious decision to accept some of those for the Earth orbital flights with the idea they were going to get fixed later going into the lunar program. In retrospect, they should've gotten fixed before the first flight and they did get fixed before Wally's flight (Apollo 7). There were a whole lot of other things that got fixed as a result of that. I'm firmly convinced, to this day, that the major success of the Apollo program in total was based on that one failure where we went back and regurgitated the whole system. We reworked everything from square one to find out where our soft spots were. We found an awful lot of them and they got fixed and we never had another major problem. Gus deserves a lot of credit for getting Apollo fixed to fly to the moon.*

Glenn: *I agree with everything everybody said about Gus. He was a good friend to all of us. Following up on what Deke and Al said, in the whole space program you're living out on the cutting edge, if you will. The margin for error isn't as great as it might be in a lot of other pursuits, such as flying airplanes. You're putting more things together so you're out there where the margins of safety are not as great as they've been. There's an old saying you still see on a plaque on the wall every once* *in a while or in a pilots' magazine, "Flying in itself is not inherently dangerous, but it is mercilessly unforgiving of human error." That fits in the time when Gus was killed. In it there were some things that never should have happened. An oxygen environment, electrical connectors, too much paper and whole bunch of things like that no one really thought were all that dangerous. Looking back, it really was. You learned by some of those errors and made a safer spacecraft and will make safer spacecraft into the future from now. Gus unfortunately had to sacrifice his life for one of those human errors that let us then step ahead to another level.*

As test pilots, you lost numerous friends in the nature of your work. Do you think there are tough luck guys? Gus had the hatch blow and then the Apollo fire.

Schirra: *I don't think we're basically fatalists; that was just statistically something that caught up to Gus.*

Shepard: *Well you know the old saying, "There are old pilots and there are bold pilots, but there are no old, bold pilots."*

Carpenter: *I think we all take the same chances. But we're all subject to human frailty, if not our own, that of others. You know what caused that tragedy on the pad with Apollo and the shuttle tragedy is a form of human frailty, and it's called complacency. We knew that Apollo machine on the pad was a bomb, pure oxygen inside and the high pressure. We knew that was not the right way to do it and we knew it was dangerous, but we'd gotten away with it so many times before and it was expensive to change. It was dangerous but it hadn't hurt us so we kept doing it. That was dead wrong and it took that tragedy to make it safe. So complacency causes great difficulty, but the end result is that it leads to progress."*

In an interview during his training in the early 1960s Gus Grissom said, *"If we die, we want people to accept it. We're in a risky business and we hope that if anything happens to us it will not delay the program. The conquest of space is worth the risk of life."*

After becoming the first American to orbit the Earth aboard Friendship 7 on February 29, 1962, John Glenn became a worldwide astronaut hero, like Russia's Yuri Gagarin. Glenn was one of the brotherhood who received the Congressional Space Medal of Honor. To those of us in the public relations world, it seemed NASA put him on a pedestal. Indeed, from the first time I met him at Marshall Space Flight Center in 1962, it was clear he was a no-nonsense astronaut who was there to soak up rocket science like a sponge.

Following his successful orbital flight, Glenn visited von Braun in Huntsville. He wanted to learn first-hand about von Braun's Saturn rockets so he could report back to the brotherhood. He witnessed Saturn's progress from concept to hardware being readied for critical testing. He visited with many of the 400,000 members of the Saturn Apollo team building one-of-a-kind facilities across the nation—from California to Texas, Mississippi to Louisiana, and all the way to Florida. Von Braun told Glenn that a test facility valued at $400 million to test the Saturn and Nova rockets was being planned for Mississippi. Glenn asked, *"Why invest that much money in Mississippi when the same facilities already exist right here in Huntsville?"* Von Braun answered, *"Stennis."* Glenn asked what he meant. Von Braun replied, *"United States Senator John Stennis, senior senator from Mississippi. We need his vote for funding and as you know John, no bucks, no Buck Rogers."*

Glenn aboard Friendship 7 in Earth orbit

Astronaut Glenn was slow to pick up on von Braun's message that day. Years later, when he became a U.S. senator from Ohio, I expect he recalled that lesson in southern politics. Vice President Johnson, who would become president after Kennedy's assassination, believed in "high tech pork." He made it clear to von Braun and the other NASA leaders, if they wanted appropriations from Congress for the space program to continue, they must share those dollars. Thus, the "NASA manned space flight crescent" emerged, covering Johnson's political friends from Florida, Alabama, Mississippi, Louisiana and Texas.

Glenn left NASA shortly thereafter and went to work for Royal Crown Cola. I remember being disappointed that one of our first astronaut heroes chose to work for a soft drink company. It didn't seem right. I thought famous space fliers like Glenn should become presidents of aerospace corporations or ambassadors to one of our allied countries—or even go into politics. Of course, he did eventually become a U.S. senator, a position he held for 24 years.

Glenn chose to run for president in 1983 and the M7 worked for him, as did many of us from all over the country. He came to Huntsville and had his kick-off event at the Space & Rocket Center. It was a bit awkward for some in the aerospace community who were staunch Republicans, but they came and showed their support for John Herschel Glenn. Interestingly, most M7 guys and many other astronauts that I've known became Republicans after JFK. In fact, many became friends of the Bush family.

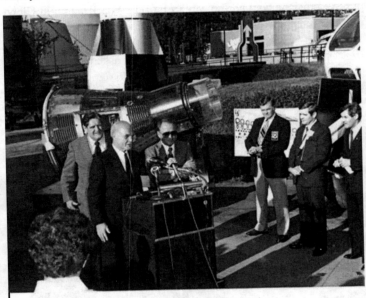

Glenn announces his candidacy for president of the U.S. in 1986 at Space & Rocket Center

One time, when Shepard and I were in Washington visiting politicians on the hill to get votes for the International Space Station, Shepard said, *"Let's go by and see John Boy."* We entered Senator Glenn's office in the Hart Building with no appointment. The first receptionist didn't know who Shepard was but the second receptionist not only knew who he was, she personally escorted us directly into "John Boy's" office. We had a nice visit discussing Space Camp, political issues, the next space mission and our families. As we were about to leave,

Shepard glanced at his watch and said, *"We have a flight in 30 minutes. John, would you mind dropping us off at National?"* Glenn replied, *"By all means, I'd be glad to."*

Glenn grabbed his coat and told his administrative assistant he was driving us to the airport. We proceeded to the special reserve parking area in the Hart Building and then headed to National Airport. Upon arrival, we thanked him for his courtesies and said our goodbyes. Impressed with the entire run of events, I turned to Shepard and exclaimed *"Man that was something!"* Shepard looked at me and said, *"Well, Buckbee, what did you expect? Didn't you know that Glenn was my backup in Mercury? Driving us to the airport, that's something John will always remember."*

At Glenn's 10th anniversary celebrated at the Cape, he talked about some of the interesting memories. After several scrubs of his Friendship 7 flight, he had to have the inevitable statement for Shorty Powers to release to the press. The first few times, Glenn told Shorty, *"Tell them anything you want to tell them."* But after he read Shorty's rendition, Glenn added, *"I decided I'd better make my own statements. I'm not sure we improved a whole lot in being profound, witty and very philosophical because there were such brilliant things that came out after that, such as, 'There'll be another day,' as if we had launched there wouldn't have been."*

Henri Landwirth, the hotel czar of Cocoa Beach who had become a friend of the M7, built a 700 pound cake that was 6 feet, 2 inches across and 9 feet, 7 inches high—the same as Glenn's spacecraft Friendship 7. I told Henri a short time ago that the most amazing thing about that cake, after he tried to keep it fresh after all those scrubs, was that he didn't give half the people in Cocoa Beach ptomaine poisoning.

At an event honoring the M7 astronauts, Glenn addressed Shepard, who served as president of the M7 foundation for 13 years, *"When we launched on the Friendship 7 flight and I finally got the go for seven orbits, and everything had been very tense up to that time, and once in orbit I'm beginning to follow the flight plan... I reached over and pull this Velcro strap off a canvas bag in Friendship 7 to get out a camera and start taking some pictures. But instead of the camera, out floats something you'll remember, (imitating Bill Dana, who played the José Jimenez character) 'You have to open the nosecone and let out the little mouse.' Remember that poor little mouse? Al had put in there a little felt mouse with pink ears and a long tail that was fastened so it would not float around the cockpit and that was really something. I had just gone into orbit and the view was terrific. I just started to work and up comes the little mouse. Al, I still have the little mouse at home and it's one of my very most prized possessions."*

Glenn considered himself to be sort of the "chairman of the board" of the M7 though not everyone thought the same way. The competition between Glenn and Shepard continued long after NASA. Shepard enjoyed the fact that he earned serious money as did Glenn. But the rivalry was friendly and in later years, I found Shepard conferring with Glenn regarding business issues that were offered the M7. When Shepard was diagnosed with a terminal blood disease, he told me Glenn had investigated research on the disease and talked with doctors all over the country trying to find a cure.

Glenn caught another ride into space, this time as a 77 year-old senior citizen aboard Space Shuttle Discovery in 1998. Some thought that ride was payback for Glenn's support of then President Bill Clinton in his last election. Others in the brotherhood thought he took someone else's seat, a younger, better qualified astronaut. But not Shepard; Shepard was proud of Glenn and let it be known that Glenn was doing it for the seniors in the country who wanted a visible role in keeping the dream alive. If José had gotten the chance, he would have climbed aboard, too.

Thirty-seven hundred press people had applied for NASA credentials to cover Glenn's launch. At the pre-launch press conferences, Glenn became a little put out that he was receiving all the questions when there were six other astronauts making up his crew. As for Glenn, he was a payload specialist seeking answers on the aging process.

At liftoff, we heard the same Scott Carpenter comment from launch control that was heard 36 years earlier at the liftoff of Friendship 7. "Godspeed John Glenn." The mission was a textbook flight and Glenn performed well for the brotherhood and didn't screw the pooch.

Back on Earth, Glenn and his crew came to Huntsville to thank the NASA Space Shuttle team and the University of Alabama in Huntsville for providing the transportation for his second ride into space and furnishing great experiments while on orbit.

In addition to receiving the Congressional Space Medal of Honor, Glenn was named among 100 outstanding aviation pioneers of the world and a charter member of the Astronaut Hall of Fame. As a Marine combat pilot, test pilot, an astronaut in two very different eras, Mercury and Space Shuttle, and a U.S. senator, John Glenn became a role model and inspiration for many.

But on February 20, 1999, it was gotcha time in Florida when Glenn's friends and the Astronaut Foundation started by Alan Shepard held a tribute attended by several hundred friends. With every tribute involving astronauts there's always a "roast and toast" and this was no different. "John Glenn— A Legend Lost in Space," was conducted by an impressive line-up of roasters; Apollo 13 astronaut Jim Lovell, fellow Mercury astronauts Schirra, Cooper and Carpenter, moonwalkers Charley Duke and Gene Cernan, Bill Dana (the José Jimenez of the '60s), and Henri Landwirth, M7's friend from Cocoa Beach days.

Walter Cronkite, of CBS-TV fame, was the master of ceremonies. He opened the evening with, *"He's John Herschel Glenn, Jr., who captured the hearts of his countrymen in 1962 when he became the first American to orbit the Earth. They cheered his name. They showered him with confetti. They named schools and roads after him. Now, after laboring in the U.S. Senate more years than the people in Ohio could count—the last two defending President Clinton—he's ready to risk his neck again….."*
And so, the games began.

Carpenter (imitating a news reporter): *This report is dedicated to a great man, John Herschel Glenn, Jr.*

Schirra: *But greatness has its drawbacks. There will also be distracters, nay-sayers, backbiters, those who will degrade and demean him.*

Gordon: *And we are all here tonight!*

Schirra: *I was asked, was I envious of John Glenn or jealous. No, I wasn't that old and I didn't need the flight time. I had 300 hours and John had 5. But I too, would do anything to get out of that crazy, U.S. Senate!*

Schirra later added that if he had been chosen to fly on the shuttle, he would not have ridden in the back; he would have been the commander.

Cooper: *Washington, D.C.—The real reason for John Glenn leaving the Senate was revealed today when an unidentified source was quoted, "No use staying in the Congress any more, all the good impeachments have been taken." I have to tell that John has always tried to impress everyone that he's on the straight and level. But ever since he won the U.S. Marine Corps Pacific Championship by doing more than 100 continuous rolls in a fighter aircraft, he's been unable pull it off. He still has trouble flying straight and level. One day early in the space program, John got up on his soapbox and said, "Men, we have to continue to convey the Boy Scout image. We have to be squeaky clean: We have to keep our zippers closed." I never did know exactly what he meant by that but it gave me a real complex. I've always had the terrible fear that my fly was open.*

In Glenn's remarks he gave it back to his friends and received some good laughs, *"This is one night to remember. You have to consider the source of the information being accurate or not. Being ridiculed by a bunch like this is like being called ugly by a bunch of frogs. But the biggest surprise is to see the most trusted man in America sink to these levels. That Walter Cronkite would fall among such evil companions as these misfits and drifters, is a tragedy."*
In reference to his flight aboard the shuttle, he said he was believed to be the first senior citizen to leave Florida in something other than a Winnebago. He added that NASA told him there would not be a walk in space for him because they were afraid he would wander off! In his closing comments he said, *"How far we have come in 37 years since Friendship 7 on this anniversary date! Tonight, I want to thank each of you for your very generous support in coming here to benefit the Astronaut Foundation which provides scholarships for higher education and Space Camp…If you talk to some of the kids who have been to Space Camp—we run into them all over the country—they are so excited. They are inspired to work together for a common goal and whether or not they are astronauts in the future, I know many may well be leaders in fields that will build this great nation of ours of tomorrow. So, on behalf of all those kids and all the people who benefit from your generosity, I want to thank all of you and (imitating Walter Cronkite) that's the way it really is, Saturday, February 20, 1999."*

Malcolm Scott Carpenter was born May 1, 1927, in Boulder, Colorado. After graduating from high school, he entered the Navy's flight training program. After his stint in the military, he enrolled in the University of Colorado where he earned a degree in aerospace engineering. He returned to the Navy, earned his wings and became a Navy test pilot. In April 1959 he entered the ranks of the M7 astronauts.

In May 1962, Carpenter completed three orbits aboard the capsule Aurora 7. He conducted experiments, photographed the launch vehicle and sun from the atmosphere-free vantage point, drifted for long periods in free flight and solved the phenomena of the "space fireflies" first observed by Glenn. Carpenter rapped on the side of the spacecraft, raising a cloud of luminous particles which turned out to be ice crystals clinging to the spacecraft. During the 4 hour and 56 minute flight, he tried to reproduce the disorientation reportedly suffered by Soviet cosmonaut Gherman Titov, later identified as the first case of space adaptation syndrome.

Carpenter inspects his spacecraft, Aurora 7

A yaw condition through retrofire consumed excessive attitude control fuel and forced Carpenter to fire his retro-rockets manually. He was successful, but late—Aurora 7 splashed down over 250 miles off course. Carpenter climbed out of the floating spacecraft and into a raft to wait for the recovery forces which arrived 40 minutes later. It was a tense moment for space watchers when CBS-TV anchor Walter Cronkite stated live, on the air, that he was afraid America "had lost an astronaut."

Knowing as you did the intense heat that John Glenn experienced during re-entry, what went through your mind about a major system failure?

Carpenter: *John's problem involved what we thought was a major system failure. Mine was a minor problem and only peripherally bothersome to me. The cabin cooling system began to work fairly well after that and wasn't a major concern and in no way related to re-entry heat problems.*

You had a fuel shortage and your automatic steering quit. How did you manage?

Carpenter: *That is a very critical time upon the reentry from any Earth orbital flight. The attitude must be held within very close limits, the retro-rocket thrust must be the right duration and the right impulse. If any of these perimeters vary far from normal it will show up as a distance from the intended landing spot. Three or maybe four failures of spacecraft systems at that critical time—if they were all relative—made me 250 miles long. It didn't bother me at all because I didn't know I had these problems at that time.*

There you were, 250 miles long, surrounded by nothing but water. What did you think about the prospects of being found out in the ocean?

Carpenter: *Again, I was not aware of the fact that no one knew where I was. I knew exactly where I was.*

Prior to the first Mercury flights, there was some concern about the effect of weightlessness on the human body. What impact did that have?

Carpenter: *In spite of all the animal work that showed dogs and monkeys could stand weightlessness, there was a vocal minority in Washington who insisted on more work being done before they would allow NASA to subject Al Shepard to five minutes of weightlessness. So we did the extra work and compiled more data to satisfy the naysayers. And in the interim, Yuri Gagarin flew, to our great consternation and Al's great disappointment, and we lost out on the very important manned spaceflight first. It didn't really impede the progress of manned spaceflight itself much, thanks to the Russian resolve. But as a nation, we didn't have the first space man.*

What do you think the Mercury program contributed?

Carpenter: *I think the most important thing to have come of all of our early work in manned spaceflight is the realization that this is a marvelous grand order of things. The experience of spaceflight is transcending. It gives you a much better feeling for your own insignificance. In today's world, it brings very closely to mind important needs of the current times, and that is the need to protect our fragile planet that needs unfortunately intensive care. Spaceflight has made us aware of that.*

Is there a connection between our willingness to do something as risky as Mercury and our willingness to commit to a Mars mission?

Carpenter: *I can't find a parallel today for the Mercury program. The Mars program is not. The reason we're now willing to take on a flight to Mars, when we wouldn't have been willing in 1961, is because so many of the unknowns have been removed. And with the removal of those unknowns comes the lessening of the perceived risk. You're afraid of what you don't know. If you know a lot, you're less afraid. That's where we are with flights to the moon and now Mars. And in my view the benefits so far outweigh the risks there is no question that it's worth doing.*

How did you deal with these risks in Mercury?

Carpenter*: We knew about the risks, but to us the benefits made the risks worth taking. Other people didn't see the benefits in the same light, nor did they see the risks in the same light. They were unwilling to risk. That's what naysayers do.*

Carpenter took a leave of absence from NASA to explore others worlds in our oceans. In the Navy's Man-in-the-Sea program, he participated in numerous dives while living and working on the ocean floor in the SeaLab habitat. He served as commander of teams who spent a total of 45 days at a depth of over 200 feet. Trainees learned to pilot a mini-sub, perform experiments inside an undersea laboratory, lock-out of a diving bell, work on the ocean floor in commercial diving gear and sleep in underwater luxury in the world's only undersea hotel, Jules Undersea Lodge. Participants received training in undersea archeology and marine ecology, exploring coral reefs and shipwrecks in the Florida Keys. Carpenter left the astronaut corps in August 1967.

Carpenter's interest in the ocean and education continued to attract him to unique programs. He worked with ocean pioneers Jacques Yves and Jean Michel Cousteau aboard research vessels *Calypso* and *Alcyone*. He was involved in NASA's Space Analog Station, a fully functional system that demonstrated concepts of space life support site maps through the use of an analog-to-space mission. The Carpenter Station was used to teach space life support systems that would one day be used on space stations, moon and Mars bases. Designed and built as part of NASA's Kennedy Space Center's Mission to America's Remarkable Students (MARS) outreach effort, the station was operated continuously for 31 days on the sea floor. Carpenter Station mission crews conducted 21 telephone links during the mission. The station linked to the MARS schools, the Johnson Space Center and the Ames Research Center. The station also made several links to a village north of the Arctic Circle. Carpenter Station's crew answered questions about space life sciences and space analogs, which increased student and public interest in the sciences and space program.

The Carpenter Station supported the NASA Challenge Project with a seven-day mission on the sea floor. The station demonstrated concepts of space life support systems to the public by modeling space missions in the undersea environment. NASA's Life Sciences organization supports human exploration in the extreme environment of space.

While inside Carpenter Station, crew members demonstrated how commitment to continuous mental and physical fitness could contribute to successful aging. The Challenge Project parallels NASA research on the relationship between aging and space flight physiological changes.

How do you compare the hardships of living in space with the hardships of living underwater?

Carpenter: *Well, living underwater is hard work when you are outside. Working against a pressurized spacesuit during a spacewalk is tough too, but it isn't done as frequently. It isn't as much a normal course of affairs as it is in SeaLab. The reason for living underwater is to work outside, and that's not the case with space flight. Underwater it's cold and dark, and even though communications have been improved, some now with helium voice unscramblers, they are still difficult. Everything is hard work. I've always felt and have said a number of times, that we outwitted the rigors of space flight with a lot of money and made it relatively easy work. And that's because we had so much money. But we don't know how to outwit the ocean. It's a very intractable foe. It's your enemy. We don't know how to outwit it, so what we really do in underwater work is overpower it with heavy iron. You just put down a whole lot of heavy steel that holds it at bay.*

An M7 astronaut and SeaLab commander, Carpenter is a dynamic pioneer of modern exploration. He has the unique distinction of being the only human being—officially—to penetrate both inner and outer space, thereby earning him the dual title aquanaut/astronaut. With his voyages to the depths of the ocean he earned the Navy's Legion of Merit. Today, Carpenter lectures on the history and future of ocean and space technologies, the impact of scientific and technological advances on human affairs, space-age perspectives, the health of planet Earth and man's continuing search for excellence.

"Thank you for what you have done for our country," said a young man to Wally Schirra while standing in a Birmingham, Alabama hotel lobby. Schirra was in town to address a 2004 Alabama Information Technology Association event and present an award to the von Braun team for bringing technology to Alabama and promoting Huntsville's "Save the Saturn Rocket" project. America, it seemed, does remember her heroes.

Walter "Wally" M. Schirra was born March 12, 1923, in Hackensack, New Jersey. Coming from a family of fliers—his dad a World War I ace and his mother a wing-walker at air shows—it was natural for Schirra to become a flier. He graduated from the U. S. Naval Academy in 1945 and from naval flight training at Pensacola Naval Air Station, Florida, in 1948. He served as a carrier-based fighter pilot and operations officer before attending the U. S. Naval Test Pilot School at Patuxent River, Maryland.

During the Korean War, he flew F-84E Thunder jets as an exchange pilot with the U.S. Air Force National Guard. He had 90 combat missions in Korea. When he was selected to be a member of the M7 crew he was a test pilot at Patuxent River.

Schirra flew on the fifth Project Mercury flight, orbiting Earth in his Sigma 7 capsule six times. He chose the name Sigma because it symbolized engineering precision, and indeed, the Sigma 7 experienced a precisely engineered flight. The splashdown was just five miles from the carrier *USS Kearsarge* in the Pacific Ocean. True to his Navy background, Schirra elected to remain aboard the capsule until it was lifted to the deck of the carrier.

Schirra and his Gemini 6 crewmate Tom Stafford participated in the first rendezvous of two spacecraft, meeting Frank Borman and Jim Lovell who manned Gemini 7. The countdown for Gemini 6 proceeded without incident and it looked and felt like a normal day at the office for Wally. Then, after ignition, the engines shut down. It was very tense, with chances of an explosion on the pad and loss of crew. Opting not to eject, Schirra gave launch personnel the opportunity to "safe" the vehicle. Feverishly going through procedures, the launch crew heard Wally utter, *"We are just lying here, breathing."* After a three-day delay, Gemini 6 launched and caught up with Gemini 7. Together, they flew formation, just a few feet away from each other, for five orbits before returning to Earth.

Schirra and von Braun talk Saturn before Schirra's Apollo 7 flight

Schirra returned to space in 1968 as commander of Apollo 7, the first mission following the devastating loss of the Apollo 1 crew commanded by Gus Grissom. During the 11 days in Earth orbit, Schirra—who would become the only Mercury astronaut to fly all three, Mercury, Gemini and Apollo spacecraft—and his crewmates Donn Eisele and Walt Cunningham successfully checked all the Apollo systems, qualifying the spacecraft for moon missions.

It had been over 20 months since NASA had attempted to fly a manned mission: The sting of losing crew and craft in the Apollo 1 launch pad blaze on January 27, 1967 continued to burn red hot. But the country had a challenge to meet—a technological prowess to prove. A lot was riding on Apollo 7 when it lifted off on October 11, 1968. The mission was considered a dress rehearsal for the 11-day moon flight.

The Apollo spacecraft had been totally re-designed under the watchful eyes of Schirra and the brotherhood. They were riding on von Braun's first manned rocket since the Mercury Redstone, the new Saturn IB. The ride on the Saturn was a real fireworks show, "like an erupting volcano," according to Schirra, but they only experienced about 3-4g's compared to almost 10 g's in the Mercury flights.

The mission was uneventful with no anomalies, NASA's favorite term when things go well. Schirra and crew carried out all of the test objectives including rendezvous with the upper stage which made Schirra, "feel like a bomber pilot." They were the first crew to broadcast live on television from space; their on-air segments became known as the "Wally, Walt and Donn Show." Their incorrigible quips included, "keep the cards and letters coming folks!"

"Keep those cards and letters coming folks," was broadcast by the "Wally, Walt and Donn show," from high above in Apollo 7

For Schirra, Apollo 7 was typical: all mission objectives met, no major glitches and a perfect landing and recovery. The only exception was the crew did not remain in the spacecraft while being hoisted aboard the carrier, *USS Essex*, as Schirra had done on previous flights.

Next was Apollo 8, referred to as "beat the Russians mission." It was the first manned flight on the Saturn V moon rocket, a mission that was undertaken much earlier than originally planned. Astronauts Frank Borman, James Lovell and William Anders journeyed to the moon (not landing) and back. The plan had been to first fly several Earth orbital missions, then go to the moon with a fly-by. But in 1968 there was concern about the Russians beating us to the moon. Dr. George E. Mueller, NASA associate administrator for manned spaceflight, proposed the mission change that was not well received by conservative rocketeers at Huntsville's Marshall Space Flight Center. After serious deliberations among senior staff, Director Wernher von Braun informed Mueller Huntsville was "go" for Apollo 8 to the moon. This first flight of an all-up Saturn V became an historic success and broke Russia's momentum.

On Christmas Eve 1968, the crew read from Genesis, "In the beginning…" It made listeners feel that we just might pull off this moon landing before the end of the decade. No doubt, these two missions—Schirra's Apollo 7 providing the first critical flight of the new spacecraft and Apollo 8 being the first manned crew to achieve trans-lunar escape velocity at 25,000 mph and loop the moon—gave us the confidence to complete the first landing of man.

I was the escort officer when Schirra accompanied a new group of astronauts on their first visit to Huntsville. Neil Armstrong, Pete Conrad, Frank Borman, Jim Lovell, Buzz Aldrin, John Young and Ed White were in the group. Wally arrived via his NASA jet and I picked him up and drove to the Redstone Officers' Club. There, for the first time, Wally saw astronaut trainee Aldrin, who had his military medals displayed on his business suit. Schirra asked, *"Who in the hell is that?"* I had to grin when I answered, *"Wally, he's with you!"*

Buckbee, left, escorts John Glenn, John Young and Frank Borman during orientation visit to Huntsville.

In 1962 the second group of astronauts make their first visit to Huntsville joined by several M7. They are from left: Jim McDivitt, Neil Armstrong, Jim Lovell, Tom Stafford, Elliot See, Pete Conrad, von Braun, Deke Slayton, Frank Borman, John Glenn, Wally Schirra, Ed White and John Young

Schirra and Shepard loved to talk about how tough things were in the old days. Blatantly chauvinistic, they would say, *"During the Mercury and Gemini days, we never had to put up with those flight stewardess astronauts."* Once, Schirra was asked by a television reporter to give his opinion regarding females being selected to fly on the shuttle. *"I don't have a problem with that,"* he commented, *"as long as they ride in the back and don't expect to fly the machine."*

During a more recent discussion about the return to flight of Columbia, Schirra said, *"I made the first flight after Apollo 1 and Rick Hauck made the first flight after Challenger went down. Now, Eileen Collins is going to fly the next flight after Columbia. I was reminded by NASA that I talked to Rick and cheered him up before his flight and it was suggested that I consider talking to Eileen before her flight."* Schirra went on to note, *"Collins is a veteran space flyer and has all the credentials for a commander. I plan to talk to her as I did Rick. I'll explain to her that I'm satisfied she should fly that mission as the commander not because she's a female, but because she's earned the seat."*

All of the Mercury astronauts and particularly Schirra, had comments regarding the movie, *The Right Stuff*. In most scenarios the book written by Tom Wolfe was close to reality. However, the brotherhood had big problems with the way Gus Grissom was portrayed in the movie. In the movie, Grissom came across as a bumbling idiot. None of the Mercury astronauts thought the portrayal was on target. In fact, they went out of their way to correct the story about Gus.

Wally wasn't necessarily pleased with his own portrayal because he thought he should've had a bigger role with his character playing practical jokes on others. Gordo Cooper was portrayed as the super duper stick and rudder man who had all the fun and masterminded the gotchas. Shepard, I think, would have preferred his character to have had more time on the screen. Overall, I think he liked his portrayal as the arrogant, swaggering Navy pilot. "John Boy" Glenn, as other referred to him, was a straight-as-an-arrow character, which Glenn liked. Most everyone agreed his character was right on. Schirra thought von Braun's German rocket team was treated poorly and misrepresented as to the role they played in the Mercury program.

Most thought Vice President Lyndon Johnson's portrayal in the movie was right on target. Schirra, however, didn't believe Johnson was given proper credit for authoring the Space Act that created the space agency. Most of the M7 respected him but weren't too happy that he led the effort to have them moved from the beaches of Virginia to hot, humid Houston, Texas. If they were going to move, they preferred Cape Canaveral or San Francisco. Schirra, on the other hand, believed Johnson, as chairman of the Space Council, made it possible for the nation to have an aggressive space program.

As Schirra tells the story, in 1962 the M7 had an opportunity to meet privately with Vice President Lyndon Johnson. NASA Headquarters and the White House thought the M7 was trying to get an additional Mercury flight. But that wasn't the case. What they really wanted was to get the Gemini program going. Johnson invited the M7 and Bob Gilruth, director the Manned Spacecraft Center to his ranch where they rode around in Lincoln convertibles chasing deer and seeing the sights. Johnson later called them into his study and informed the M7 that President Kennedy had told him to see what they had on their minds. Schirra related that Johnson picked their brains and assured them, *"he would get the Gemini program off it's ass and get going."* All of them came out of the meeting feeling good and totally impressed with Johnson.

During the first orbit of your Mercury flight, you were having the same malfunction as Scott Carpenter when your spacesuit failed. What happened?

Schirra: *We were worried about the mission being cut short because there was heat build up in the suit circuit. This was the area I was most responsible for during Mercury, the environmental control system and the suit. At any rate, I knew the system so well that I knew what the problem was. I had this little valve that I could control. What I had done was make a decision that I would turn this valve very slowly and not whack it back and forth, trying to find the right setting. So I kept watching the suit temperature, like looking at a gas gauge as it goes towards zero. In this case, I was watching the needle to see if it would stop increasing in temperature; stop and back off again. Finally, I got it under control.*

What did you like about Gemini and how did it help us get to the moon?

Schirra: *A good answer for both of those would be Gemini 6. We had to do a rendezvous but the beauty of it was that Gemini 6 didn't do its mission alone. Gemini 7 was there as our target. We had to endure in space for two weeks to do the round-trip to the moon. In fact, at one point, we considered very seriously sending the Gemini around the moon if North America* (North American Aviation, Inc. builder of the Apollo command and service modules) *didn't get their act together in time. The Apollo 7 flight for me did validate that the vehicle—not the crew—would last almost 14 days. That was satisfying, but it was very boring.*

When you rendezvoused with Gemini 7, what did you see?

Schirra: *Frank Borman and Jim Lovell had been up there 11 days when we arrived on orbit. Lovell later described their mission, "like spending fourteen days in a men's room." We chased them down and maneuvered right up to them, coming within inches at times. I noticed this long slender structure hanging down from their spacecraft and couldn't identify what it was. Turns out it was the urine dump value. After 11 days in space, the urine that was dumped overboard had frozen on to the side of the spacecraft. I called it the Gemini 7 pissicle.*

Why did space require more skill and concentration than any combat mission you were involved in?

Schirra: *I use the old cliché about combat, "Combat has within it hours and hours of boredom interspersed with moments of sheer terror." That's a common aviator creed I might add, as well. The spaceflights were practiced so much that as a result most of the actual flights were an aftermath. There* *was easy follow-up to the worst cases you practiced. NASA basically trained the fear out of us. Spaceflight is not like combat. Combat is a pretty tough environment and a good test of your capabilities.*

Tell me about the Mercury Redstone launch you witnessed before Shepard's flight.

Schirra: *My favorite story is about one of the aborted Mercury Redstone launches. It launched about one inch and settled back down on the pad. I was standing with Max Faget, who designed the Mercury spacecraft and he said he didn't think the Redstone took off that fast. I had to tell him, "That wasn't the Redstone, Max, that was the escape tower." The blockhouse decided to let the launch program run just as if the Redstone had lifted off. Next thing you knew, the chute came out and hung in the wind, sea dye marker running down the side of the Redstone looking like Mrs. Murphy's laundry.*

Mercury spacecraft designer Max Faget and Wally Schirra observed this aborted Mercury Redstone during attempted liftoff from Cape Canaveral

The other part of that story is Kurt Debus, von Braun and Bob Gilruth were in the blockhouse watching this whole sequence taking place on the Redstone pad. Debus said, "How are we going to drain the fuel out of the rocket?" Von Braun replied, "Kurt, go home and get your rifle and put some holes in it." Gilruth almost had a heart attack.

How did you know what was happening during the aborted Gemini 6 launch?

Schirra: *That was probably the toughest moment I've ever had. Fortunately, I'd had a liftoff before on Mercury and I knew what it felt like. I didn't have that on Gemini. We had simulated such things as abort, launch aborts and kill the launch. We never had to kill a launch, which was when the booster would shut down after it started. I knew what I had in milliseconds—we hadn't lifted off. Yet, all our indications were that we had lifted off. From Alan Shepard's first flight, we always said, 'We have liftoff.' I didn't say that and mission control said, "Liftoff," and the clock started. I didn't say that because we had not lifted off. That came from experience and part of it probably was combat experience. Part of it was all the years I'd trained for Mercury and all the years I trained for Gemini. It all came together in a millisecond and the decision was made.*

Wally made a number of trips to Huntsville in the '70s. He came so often we decided to "southernize" his name—"Wally Bob" Schirra. He felt more at home in Alabama with a second name. In fact, he liked it so well he asked me to add "Wally Bob" to his hat. The hat was a promotional device I developed that was referred to as the honorary astronaut hat. On the front was labeled, "Honorary Astronaut, U.S. Space Camp, Huntsville, Alabama." Shepard referred to it as Buckbee's "tacky hat."

Those hats were presented all over the world to Vice President Bush, VIPs, CEOs, heads of state, movie stars, airline pilots; you name it, if they were in the presence of Shepard, Schirra or me, they received a tacky hat.

Schirra's frequent trips to Huntsville were partly because his Mercury spacecraft Sigma 7 was on display at the Space & Rocket Center. He loved to come by on his flight anniversary, talk to the press and re-set the flight clock on the instrument panel of Sigma 7. He did several commercials at the center, one for the cold remedy Actifed and one for Tang, the "official drink of the astronauts." He spent several days at the space museum with a TV crew shooting a Tang commercial for General Foods that I don't believe ever ran. When the time came to do the final take for the commercial, we changed the label to "Prune Tang." Wally walked out in the front of the camera and held up the jar and read, "Drink the drink of the astronauts. Drink Prune Tang." That was my gotcha for Wally and he loved it.

Schirra and Buckbee at press conference with Schirra's Sigma 7 Mercury spacecraft

Schirra with kid actors making Tang commercial for Space Camp

After Challenger went down, the return to flight in the shuttle program was a big event. My buddies from Kennedy Space Center's public affairs office called to see if I could get one of the Mercury astronauts to come down for the launch. The story was they would have more press at this launch than Apollo 11, the first moon launch. I spoke to Wally and he said he could make it. We hooked up at the Orlando airport and stayed at Cocoa Beach. The next morning we went out to the press site. There was a huge press turnout from all over the world with at least 40 or 50 crews lined up with their cameras pointed toward the launch pad. Schirra did several interviews and as we were departing to go to the observation deck he did one final interview with the crew on the end of the row. Concluding the interview, he decided it was a good time to tell a few jokes. The one he told the TV crews was, *"Why was Shepard the first U.S. astronaut selected to ride a rocket?....Because NASA ran out of monkeys!"* With that we left and as we were walking away, Schirra looked back and said, *"You watch, those crews will tell that joke to the next crew and then that crew will tell it to the next and the next crew."* And sure enough, as we were walking away, Schirra took great joy in looking over his shoulder and seeing his joke being passed from one TV crew to the next. By the end of the day, I'm sure it was repeated 20 or 30 times. As we continued to walk away, Schirra said, *"That was a great gotcha, Shepard."*

Is it true that you M7 guys voted on who you thought should go on the first flight and some of you voted for yourselves?

Schirra: *Oh, no, we would never do that! My vote then and now is for Alan B. (Shepard). The only voting I can recall is peer rating, where you had to vote for someone other than yourself. I kept telling Shepard I made a mistake, I voted for him. He always got a laugh out of that. Now that I think about it, I never asked him who he voted for. That would be interesting to know.*

My guess is if Shepard couldn't vote for himself, he voted for Skyray. Their relationship was very unique. Tough competitors on one hand, and each a member of the other's fan club on the other. Always, they maintained a close relationship. The Schirras consider Laura, Shepard's oldest daughter, their "back up daughter." Shepard's wife Louise, and Wally's wife Jo, were close and dear friends.

You're the only one of the group who flew Mercury, Gemini and Apollo spacecrafts. Were you close to getting a moon shot?

Schirra: *Somewhere toward the end of Gemini, I got the feeling that the Mercury astronauts were not destined to ever fly on Apollo. That was kind of a surprise because we all thought, and we still do think, we're quite young. At that point, we were obviously young enough to make that mission. I initially was assigned as backup to Gus Grissom for that first attempt in the early Apollo spacecraft that wasn't able to go to the moon, just circle the Earth. We dropped that second block which I was initially assigned, too. I thought well, this might be the end of my Apollo career since I was backing up Gus. But for the loss of that crew, I probably would not have had an Apollo flight. I had left the program and was absolutely flabbergasted when I found out that Al had a flight going to the moon. So we changed the rules and that was appropriate because we were not that old. It might very well have been that if John had stayed he might have had a whack at it. There is no age limit at this point and time. In fact, Walter Cronkite and I have discussed about how he made the first cut on the journalists in space. He is now waiting to see whether NASA's plumbing is failing as fast as his might.*

Do you wish you'd stayed and possibly gotten a moon flight?

Schirra: *Those of us who commanded an Apollo flight did not get a second flight. The only changes in that I guess, was Tom Stafford and the Apollo-Soyuz. Quite candidly, three flights in ten years was a heck of a lot of work. Before we lost Gus Grissom, he and I were comparing notes. In one calendar year we were on the road 280 days, and that was a long, hard workout. So you can't go on doing it forever.*

Wally did a lot of interesting things after retiring from the astronaut program. One of the best contributions he made was as Walter Cronkite's sidekick on manned space flight analysis. He referred to Cronkite as "Uncle Walter." In preparation for the first moon landing, Schirra asked Cronkite what he was thinking about saying as the first human, an American astronaut, stepped on the moon. Cronkite seemed unconcerned. Just before they went on the air, Schirra again asked, *"Walter have you thought what you'll say when they land?"* No response. As the dust cleared on Tranquility Base, and the world heard Neil Armstrong transmit the now famous words, *"Houston, Tranquility Base here; the Eagle has landed,"* Cronkite came back with, *"Whew! Golly! Gee! Wow!"*

Buckbee and Armstrong in lunar module simulator Armstrong used for training on display at the museum. When given the go by Houston for launch from the moon, Aldrin said, "Understand we are number one on the runway."

Fast forwarding to late 2004 and the promise of commercial manned spaceflight and the potential of space tourism, the X-prize competition drew attention to space travel in the same way Charles Lindbergh did transatlantic flight with his non-stop to Paris. The competition dictated a team had to fly a space ship with one crew member to 62 miles altitude and successfully complete the same mission within two weeks. I asked Schirra what he thought about the 62 mile-high flights of Burt Rutan's SpaceShip One—the successful winner of the $10 million X-prize competition. Schirra's emailed response: *"An infield pop fly with a spin."* I asked him to elaborate on the subject.

Is it like the times when Chuck Yeager and his buddies were in the high desert visiting Poncho's Happy Bottom Riding Club and pulling airplanes around with pickup trucks? Will this benefit the nation, add to technology?

Schirra*: One thing I admire is the propulsion system: That's pretty good and it looks like it's pretty safe. The rest of it seems more like kid stuff, show and tell, amusement park stuff. These guys fly without pressure suits. I remember when I had to be fitted for a full pressure suit and had to take altitude training and this was above 50,000 feet, 10 miles. In the military you don't fly near 50,000 feet without a pressure suit. These guys fly without pressure suits. Then, in Mercury we flew with a complete spacesuit and we fought like hell to get ONE window. This guy flies without gloves with M&Ms floating around and windows all over the place. I can't believe how amateurish they are. I think it's safer to fire the guy at the circus out of the cannon into a safety net. They have a parachute but they don't seem to have room in this thing to wear a full pressure suit. That's just an unwarranted risk.*

Wally and others members of the brotherhood aren't happy about the SpaceShipOne pilots receiving astronaut wings for an 18 minute flight, 62 miles up. M7 and others had to train for years and fly over 100 miles up and into orbit on converted military rockets to become astronaut qualified. To them, it seems the standard has been lowered.

Mercury Astronaut L. Gordon Cooper, called Gordo by the M7, was born in the heartland of the nation, in Shawnee, Oklahoma. Like many of his astronaut colleagues, he developed an interest in flying by washing airplanes, hanging out and doing odd jobs at the local airport. His best friend—his dad—was his greatest mentor, though. The senior Cooper flew as a pilot with the Navy in WWI, and was acquainted with Wiley Post and Amelia Earhart; pretty good role models for a young aspiring aviator. Gordo was exposed to many kinds of aircraft as his father served on active duty in WWII in the Army Air Corp. Although he flew several different kinds of aircraft and was familiar with various flying maneuvers long before he was in his teens, Gordon Cooper didn't take any formal lessons until he was 15 years old. Later, he earned money by working the flight line at the airport and he spent every dime on flying lessons.

Cooper completes the last Mercury flight

Cooper graduated from the U.S. Air Force Institute of Technology and the University of Hawaii. Following flight training he flew F-84 and F-86 jets. He also attended the U.S. Air Force Experimental Flight Test School and was assigned as test pilot.

He was accepted in to the M7 program in April 1959. As a Mercury astronaut, Cooper piloted the Faith 7 capsule on the sixth and final project mission, executing 22 orbits in a 34-hour flight. Gordo nearly had a fire with a short circuit. Like the other capsules, his was packed with wire bundles that were not well protected against moisture. Near the end of the mission he had to manually fire his retro-rockets and steer the spacecraft through re-entry, overriding Faith 7's automatic stabilization and control system. With carbon dioxide levels rising in both his suit and the cabin, Cooper issued one of those classic M7 understatements, *"Things are beginning to stack up a little."* He radioed that he would bring the spacecraft through re-entry under manual control.

You were the last astronaut to fly in the Mercury program and the first American to spend a full day in space. What was your strongest impression of that first full day?

Cooper: *It certainly didn't seem long because there were a lot of new and exciting challenges and many things I wanted to do in addition to the things I had scheduled to do. I had a number of personal things I intended to fit in if I could, such as photograph the Himalayas. I was the first to do this and I felt an experienced photographer in space could come home with some good pictures. Water experiments in plastic containers and a few other things were on a time and space available basis. I would've liked for the flight to go on a little longer.*

Nineteen orbits and suddenly the flight controls go out and you're controlling the spacecraft manually. How did you feel when that last system failed and you realized you were completely on your own?

Cooper*: Well, I had trained for that in the simulator. It was one of the worst cases I can think of yet, having complete electrical failure. I felt I was ready for it and could handle it and I was the one who would get me home. If I didn't, no one else would.*

Did you really go to sleep before launch?

Cooper: *Yes, I had all my work done and we had a little hole* (in the countdown) *built in there to get all the range information. It was an opportunity to get a little rest, so I did.*

Gordon Cooper, Buckbee and Shepard at opening of Space Camp Florida

Cooper became a two-time space flier when he made his second flight as command pilot of the Gemini 5 mission. He and Pete Conrad established a space endurance record by traveling more then 3.3 million miles in 190 hours and 56 minutes. It was the first spacecraft to use fuel cells. The fuel cells behaved erratically, requiring flight controllers, and the crew had to nurse them through the whole mission. Cooper and Conrad spent the last days of their flight in a "powered down" drift to conserve electricity.

Cooper felt the eight days in Gemini were more trying than the Mercury flight, noting the Gemini flight felt like eight days flying in a garbage can. He added the Gemini mission had more than it's fair share of problems.

Often, in public forums with his colleagues present, Cooper would be asked who the best pilot in the bunch was and Cooper would answer in that Oklahoma drawl, *"You're looking at him."*

Before he left NASA, one of his jobs was to review training for the Skylab program. He came to Huntsville to review Skylab equipment being designed for use by the astronaut crews. He was very critical of many of the Skylab handholds, foot restraints and hatches as, *"being designed by people who didn't have an appreciation for zero g."*

After he retired from NASA and was working for Walt Disney Enterprises, I visited with Gordo to discuss the idea of Space Camp for kids. He loved the idea and said, *"We should be doing that here at Disney.*" He became a big supporter and was present along with Alan Shepard when Space Camp Florida was announced.

Gordo was portrayed in *The Right Stuff* movie as rambunctious, rowdy; a fly-boy that let it all hang out. I think he liked that, even though that wasn't the Gordo I knew. Cooper was more analytical than emotional. As Schirra said, *"He was more of an engineer than a test pilot."* He liked to think that technology was there to make a better engine, to improve the lift of the aircraft, to find an alternative fuel. He was always searching for a new and better ride.

Cooper, the youngest and the last astronaut to fly in a Mercury spacecraft, died October 4, 2004 at the age of 77.

The three remaining members of the M7 brotherhood—-Scott Carpenter, John Glenn and Wally Schirra—gave Gordon Cooper a fighter pilot/astronaut tribute in the presence of his family and close friends at NASA's Johnson Space Center on October 15, 2004.

He was often described by his peers as "the best damn stick and rudder man in the bunch." Glenn shares with the audience his memories of Gordo. *"A thousands memories come out when something like this happens. The only time I saw Gordo speechless was after a fishing experience he had at the Cape. He had gone to a large lake near where we were staying, wading in his swimming suit and fly fishing. When he returned he said, "I saw the biggest, dang bullfrogs out there in that lake I ever saw in my life."* A guy who lived at the Cape stood up and said, *"Were you fishing from the bank?"* Gordo said, *"No, I was wading about up to my waist."* The local guy said, *"Those weren't bullfrogs, those were alligators!"* *"It's not a wake, it's a celebration of Gordo,"* continued Glenn. At the end of his mission during re-entry, Gordo used the term, "right on the old gazoo" as Glenn gave him a countdown for retro-fire and he manually lined up his Faith 7 spaceship for re-entry and recovery. What you saw is what you got. *"You could always depend on Gordo,"* said Glenn.

"Nearly 50 years ago, a small group of American men were given a special charge by this nation to ensure pre-eminence in space. This was a time when world opinion held that pre-eminence in space was a condition of national survival," recalled Carpenter. *"With that charge, this group that came to be known as the M7 began to compete among themselves to see who would get to take the first flights to guarantee our national survival. That competition added to the group's solidarity and the strength of our fraternity. We were welded into a fraternity that had no equal at the time. Time has diminished our group. We are no longer seven. We are three. The Soviets are not twelve but five. We are here to mark the passing of one of our number, Gordon Cooper. We mourn his passing. Gordo's contribution was essential to the group's solidarity and we celebrate his contribution and remind ourselves that nothing is in the constrict of man. It is proper to say farewell, Gordon Cooper. It was an honor to be a member of your fraternity,"* concluded Carpenter.

Recalling the order in which the M7 were always called on stage, with the abbreviation of 'CCGGSSS' —Carpenter, Cooper, Glenn, Grissom, Schirra, Shepard and Slayton—*"Now we are out of order and I'm the only smart 'S' left,"* said Schirra. He added, *"We regret losing Gordo. He was one of our dear friends, not a bad water-skier, not a bad pilot, but a heckuva good astronaut. We are probably the most bonded seven men in the history of aviation and space and mankind, and to lose another one is pretty tough for us,"* Schirra finished.

"Many people are more inspired to do great things because of his adventures in space," said Henri Landwirth, one of Cooper and M7's long time friends. Landwirth, founder of "Give Kids the World," described Cooper's sincere interest in helping the terminally ill kids who visited.

Sean O'Keefe, administrator of NASA, closed the ceremony with a presentation to Suzie Cooper of the NASA Distinguished Service Medal, saying,

"In my youth these were the super heroes of our time…We asked the best and bravest to help lead the way in this new ocean of space. We members of the NASA family are forever grateful that Gordon Cooper answered America's call."

Glenn said, *"In flying terms, most of the people here have a lot more runway behind them than ahead of them. In fighter-pilot terminology, Gordo has scrambled. He's out a little ahead of us with Gus, Al and Deke. I'm sure we will all rendezvous out there someday."*

Linn LeBlanc, Executive Director of the Astronaut Scholarship Foundation, remembers one of the best "gotchas" in years.

"Gordon & Suzi Cooper, Scott & Patty Carpenter, Al and Jill Worden, Bill and Evy Dana and I all boarded the American Queen on Monday for the Astronaut Scholarship Foundation fundraiser 'Astronauts on the Mississippi' cruise. This was a five-day riverboat cruise, out of New Orleans, with a three-day cruise immediately following. Wally was to join the group at one of the Port of Calls on Wednesday and stay for half of both cruises.

"So while the cats away...the astronauts and spouses started thinking of a proper welcome for Wally. 'It has to be a gotcha...and a good one,' I remember Cooper saying. 'And, we have to be there to see it,' Worden added. He went on to explain whenever they would pull a 'gotcha' on Wally in the past, he'd never acknowledge it. He told of the time Wally called Jim Rathman to order a new, specialized tire for his corvette, that you could only get overseas. Al and some of the others, pull a tire out of a junk yard, spray-painted the spokes and rim and sent it up to him. They know Wally received it, but never said a word.

"So the 'gotcha' plan ranged from having Suzie in his room with another man, to having him sit alone at his meals. We finally came up with a plan—have the hotel manager, Tony Suttile, in Wally's bed with one of his woman staff members. They both were to dress down to their skimpies and be in a compromising position. At the day's end, we had still not solidified the room or time, but in the morning the hotel manager placed a flyer under everyone's door labeled 'Operation Wally.'

"To give you an idea of the room layout, when you first enter, there is a small hallway with a bathroom to your right and a closet to your left. At the end of the hallway, the room opens up and there are two twin beds to your right. The set up was perfect—a suitcase on the closest bed to the hallway and champagne and strawberries on the dresser. Tony and the woman staged in the far bed lying with Tony's back to the door and the woman on the other side of him. Now that the stage was set, we were ready to roll. I went down to the port to meet Wally, while the other astronauts and spouses waited in the bathroom of Wally's room. Remember, they all wanted to be there when it happen.

"Well, we had planned for a fifteen minute window to get Wally up to the room—it took forty minutes. I heard later the bathroom was quite a joke in itself. There wasn't room for everyone, so Gordo and Suzie stood in the bathtub; Patty Carpenter sat on the toilet and the others stood shoulder to shoulder in the center. Patty decided that Wally might see the light from under the door, so they decided to turn it off. Thus, they sat in the dark.

"So I meet Wally and get him up the room. The door was unlocked, which Wally thought was strange, and we walked in. We turn the corner into the bedroom and this woman lets out a horrific, high pitch scream, while laying in the bed with Tony. Wally must have jumped two feet in the air, his mouth wide open, he looks at me and then back and them. Meanwhile, Tony, the hotel manager, jumps out of the bed in his boxers and tank top and yells to Wally, 'Who the Hell are you!?' (I may not have mentioned that Tony is easily 6 ft. 4 and towered over Wally.)

"Wally, still with mouth open, stutters '...your door was unlocked.' Then Tony says, 'Wait a minute. You're Captain Schirra, aren't you?'

"Wally nods (mouth still open). 'I have your book...will you sign it?'

"Wally looks at me and laughs (still a bit startled), patting his pockets for a pen. Once he finds it, Wally opens the book and the first page has a big sign that reads 'GOTCHA'. The astronauts and spouses came out of the bathroom, clapping and hollering. Wally later said that was the best Gotcha since Al Shepard. He came down to breakfast and says to Gordo, 'I don't even have Parkinson's and I'm still shaking 30 minutes later.'"

Donald Kent Slayton, "Deke" to the M7, was born March 1, 1924 in Sparta, Wisconsin. At the age of 18 he enlisted in the U.S. Army Air Corps and earned his pilot's wings a year later. As a B-25 bomber pilot he flew 63 combat missions in WWII. After the war, he received his aeronautical engineering degree from the University of Minnesota. While working for The Boeing Company he was recalled to active duty during the Korean War. Slayton later became a test pilot for the U.S. Air Force and was selected as a Mercury astronaut in 1959.

To get to the top of the pyramid you had to be selected. There has always been a bit of a mystery as to who and why astronauts were selected for particular flights, but you can be sure that Donald 'Deke' Slayton, the "godfather of the astronaut corps," made the selections that were eventually approved by NASA headquarters, including who would command the flight, who would walk on the moon and who would stay behind in the command module. His fellow Mercury astronaut, Alan Shepard, assisted Deke in the selection processes. Interestingly enough, both were on non-flight status because of medical problems when they picked their buddies to go the moon.

At the time Mercury was about to go, just about everything was failing. About the only thing that didn't fail was men. Is that what you were setting out to prove, you couldn't replace man?

Slayton: *No, I don't think so. Those of us who were flying had no question about man's ability to produce. The real issue was whether the hardware could be made to produce.*

When the Mercury program ended, did you think you were winning or losing the race to the moon with Russia?

Slayton: *At that time I thought we were losing. We were behind them in every area you could imagine. They were doing extravehicular activities(EVA), multi-manned things and had more hours in space than we did. It occurred to me that we were rather far behind at that point.*

You were scheduled to fly on a Redstone but were taken off flight status. Do you have any idea as to why NASA replaced you?

Slayton: *I think it was super conservatism. Bill Douglas, our flight surgeon, who is also a member of the Mercury 7 Foundation, convinced everybody in the Air Force that I was clear to fly. After John Glenn's flight, the top-level management in Washington started looking at my situation and discovered I had this irregular heart rate. Based on that, they jerked me off. It took 10 years of re-evaluation before I finally got back on flight status.*

When Deke was grounded, the M7 had one of their séances to privately discuss Deke's situation. M7 agreed they needed someone in charge of the astronaut office that they could trust and had confidence in—not some admiral or general put in charge of their future who knew nothing of their quest to go higher and faster. Deke took that job and made it an office with prestige and power. Nobody bucked the final say of his astronaut office. After Deke was assigned the Apollo-Soyuz mission, other people headed the office with less clout, and so the influence and power went away.

You said you had many roles in manned space flight. Is there one that stands out?

Slayton: *Well, I was primarily manager of the astronaut corps from that 10-year period from*

Mercury up through the Gemini and Skylab. I don't remember any one situation that was more dramatic than others, but Apollo 13 ranked high. I think each flight, whether it was Mercury, Gemini, Apollo or whatever, stands alone. Each one was very important at that point. Fame was fleeting and it lasted about as long as it took to get to the next one, and then that one became the most important.

You kept up your fight to get into space for a long time. What kept you going?

Slayton: *I never really thought I wasn't going to fly. I just didn't think it was going to take 10 years to prove it. I was always convinced I was right and was going to prove it sooner or later. I was hoping it would be sooner rather than later.*

He finally got his chance to fly, 16 years after giving his buddies in the brotherhood the ultimate ride. Slayton was assigned the Apollo-Soyuz mission which involved docking an Apollo craft leftover from the moon landing program, with a Russian Soyuz craft while in Earth orbit. A few years earlier, he might have been strapped in a single seat, extreme-range U.S. fighter plane with bombs onboard headed for Moscow. But instead, Deke Slayton was about to join some other fighter pilots from the "evil empire" in space.

Deke, the old man at 50 years of age, was joined by Tom Stafford, the most experienced space flier of the group with three previous missions. Rookie Vance Brand was the third member of the U.S. crew.

Deke Slayton, left, got his ride with Tom Stafford, Vance Brand and Russians cosmonauts Alexei Leonov and Valery Kubasov aboard the Apollo- Soyuz spacecraft.

The Russians named Alexei Leonov, the first human to walk in space, as the commander of the Soyuz. Leonov would be accompanied by engineer Valeri Kubasov. Slayton soon learned that Leonov was as crazy as most fighter pilots, so he fit right in with the brotherhood. He was a fun-loving extrovert who appreciated gotchas, too. Learning each other's language was a greater hurdle than many of the technical problems. For years, Deke's "vast vocabulary of Russian" was a joke with Shepard and the brotherhood. Stafford was a little better, but when you added the Oklahoma accent, it was certainly less than perfect. Officially there were only two languages, Russian and English, but the Soviets claim that Stafford's heavily accented Russian comprised a separate language, "Oklahomski." And a fourth language, developed from joint slang, was called "Ruston" (Russian plus Houston).

Deke and crew launched July 15, 1975, propelled by Huntsville's Saturn IB, the last Saturn to fly men into space. The Russians flew on their old reliable Vostok. The Apollo and Soyuz completed a rendezvous two days later and performed the first docking of two manned spacecraft from different countries. The mission was successful in perfecting a common docking device that could be used on future flights. While the Cold War was temporarily put on hold, three Americans and two Russians met in space and had one hell of a party. No doubt, the Tang was laced with a little vodka!

As you look back at it, what stands out about Apollo-Soyuz?

Slayton: *Well, at that time we had high hopes that this was going to create international cooperation, particularly between us and the Russians because that was the purpose of it. Ultimately, as everybody knows, our relationship went downhill fairly rapidly three or four years later and that was a big disappointment. I think we're trying to recreate that now, talking about going to Mars as a joint mission. To me it makes a lot of sense. I would hope that what we created in Apollo-Soyuz would form a basis for going on to Mars in joint missions.*

How did it feel to meet cosmonauts face-to-face?

Slayton: *Well, it was a pleasure to have an opportunity to meet anybody in space. I spent so many years trying to get there, I wasn't fussy about who I went with or who I met: I was just happy to be there. They turned out to be very delightful people to work with. In any case, we had no problem at all with the relationship. On a personal basis, they were really neat guys.*

How did you deal with the attention from the public and the press?

Slayton: *That was probably my biggest shock. When we were thrown before the cameras, a bunch of military pilots, and suddenly we were in view of the whole world and most of us didn't really know how to cope with that. Actually, I had a tough time with it for a long time. At least I learned that most of the press people are trying to do a job and more likely to help you than hurt you. So, a lot of them got to be good friends. That's the way it should be.*

Let's talk about crew assignments. How did you make selections?

Slayton: *We had to have someone in charge of the crews. Al and I made the crew selections and normally assigned who would be the commander in charge. Most of the time, it was really the guy with the most experience, a fairly easy decision.*

When a commander was named to be on a flight it meant he was the guy calling the shots. This was a carryover from the military where the captain of the ship is truly in charge. Commanders often selected their flying partners, were in control of training activities, were introduced first, took all the plum assignments and really were the star of the mission. In regard to the moon flights, they decided who would walk on the moon first on their flights and wore red stripes on their spacesuits to denote commander. It didn't make any difference what they had done or where they had been. When they were on the moon, the commander was also in charge of the spacecraft, including the moon buggy. In most cases, he was the only one who drove the lunar surface vehicle. In Neil Armstrong's case, the reason you see Buzz in most pictures, is because Neil was in charge of the camera. When they returned from their flights they were again in charge and made decisions as to where they went, what they did on worldwide tours and were always introduced first as the commander.

I experienced the commander rule, firsthand, when Shepard and I were in Japan announcing the opening of Space Camp Japan. Donn Eisele, Wally Schirra's command module pilot on Apollo 7, was also there, working as a Space Camp consultant. Shepard and I were called early one morning and informed that Donn had died from an apparent heart attack. Shepard visited with the authorities and confirmed Eisele's identity and worked with the U.S. Embassy to make sure everything was in order. One of the first things he did was to call Donn Eisele's commander, Wally Schirra, so Wally could notify his next of kin. He was trained by the Navy and NASA that you should inform the commander of the incident and let him take it from there. He believed that was the way things were done. I also suspected Shepard was like many of his buddies and didn't like to get too close to death.

Another question about commanders came up when the M7 were discussing assignments for moon missions. Schirra indicated that once you were named a commander of an Apollo mission, you didn't get another mission, which explained why Schirra didn't get a moon landing mission.

"We never had a rule like that," objected Slayton.

What were the considerations in deciding who made a flight. Did public relations come into play when making flight selections?

Slayton: *Well, that's a very complicated subject but public relations aspects were not major considerations. From my point of view, I think the most important thing was to have highly competent guys who could do the job technically. We talked about sending a chemist to the moon because we could get a lot of great data back. But I used to remind folks that a dead chemist on the moon didn't do any of us any good. The name of the game was to get somebody up there and do the best we could with them and get them home again.*

Were there any suggestions or outside pressures indicating one candidate would be better than another on television or other media?

Slayton: *No, never had any of that kind of pressure at all. We had some outside influences exerted once in awhile to try and get people into the astronaut corp. We never had any pressure on crew assignments. When Al was in the chief astronaut office and I was director, he and I worked together on it. I don't know of any case when he and I were second guessed or overridden.*

I worked with Deke on several public appearances that took us to the Cape and the Paris Air Show. He was the real deal, nothing made up or put on. What you saw was a true fighter pilot who just happened to be an astronaut. He always told it like it was. One of his pet peeves was references to the guys as "former astronauts." Once, when I was introducing him, he told me in a manner only Deke could deliver, *"Don't refer to us as former astronauts. Once you're a pilot, you're always a pilot. Same is true of astronauts. I can't stand to have guys like Al, John, Wally and Neil called former astronauts. It sounds like they have been stripped of their rank, of their accomplishments. It's another one of those NASA rules that some headquarters-type came up with that makes no sense."* I stopped referring to any of the astronauts as former astronauts.

Slayton, one of the first to be honored at the Astronaut Hall of Fame, was a frequent visitor, who participated in the induction of his fellow Gemini space fliers. He was an inspiration to the youngsters attending Space Camp Florida and worked tirelessly in the M7 Foundation's efforts to raise scholarship funds for science students. Slayton was president of Space Services, Inc. , a Houston subsidiary of EER Systems, Inc. He became a rocket man, leading a company in the development of rockets for suborbital and orbital launch of small commercial payloads. An avid pilot, Slayton flew a variety of piston powered aircraft until he became ill. Slayton died in 1993, the second member of M7 to be lost after Gus Grissom.

On July 20, 1969, a man from Earth stepped onto a large, dusty rock in space to claim it for America. Eleven other men eventually followed the first, and they came to be known as moonwalkers.

Since Columbus first stepped onto the shores of North America, Americans have cherished a spirit of human exploration. It was President John F. Kennedy who put into motion the extraordinary exploration of the moon. The spirit was never more evident than on September 12, 1962, when Kennedy told a group of students at Rice University, *"Many years ago, the great British explorer George Mallory, who was to die on Mount Everest, was asked why did he want to climb it. He said, 'Because it is there.' Well, space is there, and we're going to climb it, and the moon and the planets are there, and new hopes for knowledge and peace are there. And, therefore, as we set sail, we ask God's blessing on the most hazardous and dangerous and greatest adventure on which man has ever embarked."*

Indeed, America did embark on a journey to the moon, and between 1969 and 1972, 12 Americans left their footprints there.

Neil Armstrong, the first to step on lunar soil, was quiet, confident and to some, the best all-around pilot in the astronaut corps. It has been said he had "ice water in his veins." He flew combat missions for the U. S. Navy, piloted the X-15 rocket plane to the edge of space, brought the Gemini 8 home after a near disaster in space and ejected to save himself from a failing lunar landing test vehicle here on Earth. He flew the lunar module to the moon's surface and successfully landed it with only enough gas left for 15 seconds of flight. At a reunion of Navy astronauts, he commented that his laptop computer had more power than the computer that managed the systems aboard the Apollo 11 lunar module. And he added his travel pay for the trip was a mere $49.10, because government food and lodging were provided.

Neil makes limited public appearances. He is the focal point wherever he goes, including the memorial service held for his colleague, Alan Shepard, at NASA's Johnson Space Center. I'm sure it's disconcerting to show up for an event, expecting the only attendees to be your friends and instead having a mob. I've seen people at events pushing and pressing so hard to meet Neil and get his autograph, that I've become fearful for his safety. He doesn't attend gatherings that have profit motives, nor does he show up to be singularly honored. He emphasizes that all of the Apollo astronauts contributed to the landings, his included, not just the guys who walked on the moon.

One of the most memorable events during my 25-year tenure as the executive director of the U.S. Space & Rocket Center and U.S. Space Camp in Huntsville, was the visit of the Apollo 11 crew for the 20th anniversary of the first moon landing. At one of the events held in the crew's honor, Armstrong said, *"The machine (Saturn V) is man-made and hence, imperfect. No one has ever made a perfect machine. But those of you who built Saturn V; came about as close as history ever recorded. We all had flown rockets, but this time we were going to ride on no hand-me-down military rocket or popcorn firecracker. We were going to ride the Saturn V, a fire-spitting, thunder-clapping, go-to-the-moon machine. It was a great ride. Thanks so much for one great ride."*

If any moonwalker has a complaint about going to the moon, it has to be Buzz Aldrin. He was a member of the first crew to land on the moon and he became the second human to walk on the moon. According to Alan Shepard, Aldrin was second because the lunar module hatch opened toward Buzz who stood on the right side in the landing craft—which meant that Neil had to exit the cramped cockpit first. I don't think Buzz has ever quite accepted that explanation. If I were Buzz, I would be irked too, being constantly introduced as "the second man to walk on the moon," followed by questions 30 years later of, *"Why were you number two?"* They did land on the moon in the spacecraft at the same time

Buckbee, left, talking with Mr. & Mrs. Buzz Aldrin, Mike Collins, J.R. Thompson, director Marshall Space Flight Center and Neil Armstrong at 20th anniversary of Apollo 11 celebration held at the museum in Huntsville in 1989

Von Braun observes from the blockhouse

Saturn V liftoff of Apollo 11

and were the first two humans to set foot on the moon. Buzz is referred to as Dr. Rendezvous. His superior knowledge and skill of rendezvous and docking techniques, a procedure necessary to successfully land on the moon, may have made him one of the best prepared astronauts to fly to the moon and back.

While the two were climbing back in the lunar module after an EVA, one of their backpacks bumped the ascent engine arming switch, breaking it off. Buzz improvised by using a Fisher space pen to reach the metal portion inside the broken switch. It worked and Buzz and Neil successfully completed the first launch from the moon's surface.

Today, Buzz is the most active astronaut from the Apollo era. He is a sought-after speaker, he has authored books and articles, and supports private sector rocket development, space tourism and interplanetary space travel. He feels an obligation to make himself available to the American public whenever possible. He believes he has a duty to not only share his experiences made possible by the U.S. manned space flight program, but also to represent all the astronauts who contributed to the flight, including some who chose not to be involved in public events.

The third member of the Apollo 11 crew was Mike Collins, pilot of the command module Columbia. Collins was known as one of the best writers in the corps and proved it by authoring several great space books. After leaving the corps, he helped build one of the greatest aviation and space museums in the world, the National Air & Space Museum in Washington, D.C., which opened in 1976 on his watch.

American astronauts Armstrong and Aldrin take the first steps on the moon in 1969: U.S. - 1, USSR - 0

The first time I met Pete Conrad—who commanded the second moon landing, Apollo 12—he was smoking a stogie and telling explicit jokes about his astronaut buddies from the Air Force. He was another one of those astronauts who ejected from a failing aircraft and lived to tell funny stories about it. Conrad was the jokester of the moonwalkers' group and collaborated with Tom Wolfe on *The Right Stuff* book. A fierce competitor, one did not have to be around Conrad very long to see he was going somewhere in the astronaut corps.

The command module pilot of Apollo 12 was Dick Gordon. Call signs were *Yankee Clipper* for the command ship and *Intrepid* for the lunar module. Conrad and crew lifted off aboard the Saturn V which was immediately struck by lightning; an anomaly promptly denied by NASA public affairs types at the time. Bean said, *"The spacecraft instrument panel was like Christmas tree lights. All kinds of warning buzzers were going off and switches had to be reset, but that old Saturn just kept on flying, thank goodness."* One engineer said it just couldn't happen. At that point, no one had asked Conrad. When they did, Conrad's Navy vocabulary kicked in and there wasn't much usable by either the print or broadcast media, other than, *"A lightning bolt hit that big mother."*

When Conrad stepped on the moon he cried, *"Whoopee! Man, that may have been a small step for Neil, but that's one long one for me!"* While he was there, he had to activate the camera that sent back pictures to Earth. Unfortunately the camera was accidentally pointed into the sun which overloaded the sensitive circuitry and burned out the picture elements. Later Bean tried to restore it by hitting it with a special issue lunar rock hammer, but to no avail. Conrad muttered, *"Take that, you low bid supplier."*

With no transportation system on the moon, Conrad did a lot of walking to locate the Surveyor spacecraft that landed two and a half years earlier. He is also rumored to have tested the mobility of the fully pressurized moon suit by doing the first and only back flip on the moon. Knowing Conrad, he probably tried. With two Gemini missions to Earth orbit, one moonwalk mission and command of the 28-day mission aboard Skylab, Conrad stood close to the top of the pyramid.

During Skylab, Conrad and crewmates Joe Kerwin and Paul Weitz had to repair a crippled Skylab space station. An often forgotten program, Skylab would become America's outpost in space for three crews during the 1973-74 timeframe. After their return to Earth, during a visit to Huntsville to thank the Saturn and Skylab workers for their successful flight, I presented Conrad and his crew a gotcha; a multi-purpose, fix-it tool which included a plumbers helper, duct tape, hammer and oversized wrench all connected and labeled, SkylabSuperTool.

Buckbee presents SKYLABSUPERTOOL to Pete Conrad on his visit to Space & Rocket Center after successful Apollo 12 and Skylab missions

Buckbee and wife Gayle, visit with Alan Bean, Apollo 12 moonwalker and actors Richard Bryant and Chico Perryman who portrayed Alan Bean and Pete Conrad in the re-enactment of the Apollo 12 moon landing mission

Conrad knew Alan Bean in the Navy and selected him as co-pilot to land on the moon. Bean carried out his role superbly throughout the Apollo 12 mission. Bean was reminiscent of the kid next door who would cut the neighbor's grass, deliver newspapers and hang out at the local drugstore. He was dependable, quiet and a good listener.

Bean took art lessons in the early days, which Conrad and others in the fighter-jock/astronaut community found amusing. Today, Bean is the only moonwalker who can pictorially re-live landing and walking on the moon. He paints the scenes on the moon he remembers, as well as what his fellow astronauts remember. He enjoys replicating those fascinating footprints that he and his buddies left on the moon. Bean also loves to speak all over the country about his experiences and has been a frequent guest astronaut at Huntsville's Space Camp. In addition to being the fourth man to walk on the moon in 1973, Bean became the commander of a 59-day mission aboard Skylab, joined by crew members Owen Garriott and Jack Lousma.

Apollo 13 was not well attended by the media. It was assumed that Americans had lost interest in the moon missions after two successful landings.

Jim Lovell, a veteran of the 14-day, Gemini 7 mission and one of the Apollo 8 guys who was the first to loop the moon, was commander of Apollo 13. Jack Swigert was pilot of the command module, call sign *Odyssey*. The lunar module pilot was Fred Haise, call sign *Aquarius*.

I was working in Houston for this mission. The launch went well and the mission was a sleeper from a public relations standpoint until, that is, the downlink in mission control came over loud and clear: *"Houston, we have a problem."* An explosion had occurred onboard the moon-bound spacecraft. It appeared the astronauts were stranded in space. What followed over the next several days was the first real-time rescue performed by NASA. It had more drama than one could have imagined. *"Failure is not an option,"* said Gene Kranz, the head flight director of Apollo 13. Miraculously, NASA know-how prevailed and the crew returned safely to Earth.

Considered a failure in the minds of NASA senior management, the crew was ignored for years as having made no serious contribution to the actual moon landings. Lovell and his crew were often left out during NASA anniversary observances because they failed to land on the moon. In the NASA public affairs world, it wasn't proper to invite Lovell to moonwalker events.

Enter Hollywood, with Tom Hanks and Ron Howard producing their version of Apollo 13, another *Right Stuff* adventure, with the now-famous movie *Apollo 13*. Overnight, Lovell, Haise and Swigert publicly became the heroes they had always been. America became aware of how close we came to losing three brave astronauts. Gene Kranz became the icon of flight directors. If they ever have a flight directors hall of fame, Kranz will no doubt be the first inducted.

Alan Shepard was the commander of Apollo 14, the "crew of three rookies". None had orbited the Earth. A lot of guys in the corps believed Shepard would have been picked to be the first to walk on the moon had it not been for the inner ear problem that temporarily grounded him.

I reminded him once that if it hadn't been for Huntsville's Redstone and Saturn rockets, Shepard would be a nobody. He looked at me with that Shepard grin and said, *"Right, and if I had a laptop computer with me aboard that Redstone, I would have gone to Mars!"*

Shepard and Buckbee hit the road selling Space Camp

At the 25th anniversary of Apollo 11 celebration, held at the U.S. Astronaut Hall of Fame, attended by space workers from Kennedy Space Center, all of the Apollo crews were introduced and asked to comment. Speaking for the Apollo 14 crew Shepard said, *"Well, they saved the best looking, the most articulate and most modest until last. Some of you may not know that the crew of Apollo 14 received absolutely no recognition within the astronaut office. Stu hadn't flown before. Ed hadn't flown before and nobody gave me any credit for 16 minutes of sub-orbital flight. So it was a crew of three rookies led by the icy commander. All of you in this room know that Apollo 14 was the by far the best of all six missions to land on the moon. I had the privilege of selecting these two gentlemen who would go to the moon with me. Most of you know that I'm the oldest man to walk on the moon and I emphasize walking because all those young guys who followed us got to drive around up there in automobiles. Let me say how important it is to acknowledge the contributions of thousands of people like you who made our missions successful and the accolades should go to you and speaking on behalf of Ed and Stu, I thank you again and again."*

Shepard retired in 1974 from both NASA and the Navy (he had achieved the rank of admiral) and successfully went into private business. For two decades, he worked tirelessly on behalf of the Mercury 7 Foundation, U.S. Space Camp and the U.S. Astronaut Hall of Fame.

Edgar Mitchell was the most unlikely guy to fly to the moon with Alan Shepard. You would have to know them both to appreciate the thought. I once asked Shepard, *"Why Mitchell? He hadn't flown anything in space!"* Shepard's response was, *"I picked him because he knew more than anyone else, including me, about navigation and getting us to the moon and back."* Mitchell also conducted a few extrasensory perception (ESP) experiments during the mission that caused the public affairs folks to answer some new and different queries about the objectives of Apollo missions. Later, I asked Shepard if he had been aware of Mitchell's ESP experiments. I had the distinct impression it was not something the two had discussed prior to the mission.

Shepard had a surprise for us all. Near the end of the 33½ hour lunar visit, Shepard revealed a specially made six iron, which produced comments from mission control, *"What's he doing?"* He quickly attached it to a lunar scoop handle, then retrieved two stowaway golf balls for what he termed a "sand shot." The 2 feet of handle and stiffness of the spacesuit caused considerable trouble on the first shot and prompted Mitchell to remark, *"You got more dirt than ball that time."* Shepard then took another drop and swung again. The second ball traveled about 200 yards, and Shepard made his famous jest that he had hit the ball for "miles and miles and miles." Comedian Bob Hope, a good friend, had given Shepard the idea for the golf shot.

Shepard with Bob Hope and Jack Kemp at 30th anniversary of Shepard's flight. Hope: "Watch out playing golf with Shepard, he counts backwards."

When asked in the 1990s would the nation ever return to the moon, Shepard had this to say, *"I think we are going to go back to the moon, no question about it. It may very well be that it's going to take research on using helium 3 and nuclear fusion processes to create electrical energy for this country. It may have to go that far, but if it goes that far that will be a good reason for going back. There's a lot of helium-3 up there—as well as a couple of golf balls."*

Stu Roosa, another rookie, was the pilot of the command module *Kitty Hawk*. Shepard and Mitchell flew *Antares* the lunar module. In Shepard's opinion, Roosa was the best qualified candidate in the command module systems. He picked people he believed would be loyal to him, would support the mission and go all out to make it successful. I never heard a negative comment from Mitchell or Roosa regarding Shepard. I think they both were grateful to have been selected by him to make a moon landing flight.

Today, Mitchell is the only living crew member of Apollo 14. Stuart Roosa died in 1997 and Shepard passed away in 1998. Mitchell consults, lectures and has visited Huntsville a number of times to attend anniversary celebrations and address U. S. Space Camp trainees.

Apollo 14 Commander Alan B. Shepard, Jr. at Fra Mauro with the U.S. flag where he and Ed Mitchell spent nine hours exploring the surface

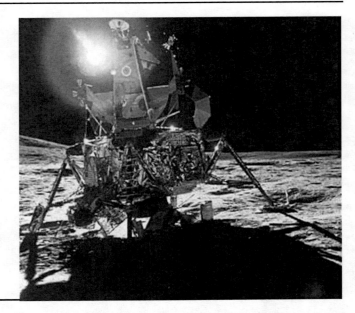

Shepard's landing craft, Antares at Fra Mauro on the moon

Shepard's out the window view of Roosa coming over the hill in Kitty Hawk

Roosa, Shepard and Mitchell, the last moon walkers to be quarantined on their return to Earth, are seen at the window of the Q van with a sign near by which read, "Please Don't Feed the Animals."

During Apollo 15, Huntsville was again front and center in the Apollo program. This was the first mission to deploy Wernher von Braun's lunar roving vehicle, another transportation system he had promised the astronaut corps. Designed and developed by Huntsville's Marshall Center, the rover became the sports car of the moonwalkers.

Dave Scott served as the commander of the fourth moon landing mission. James Irwin was the lunar module pilot who joined him on the moon and Al Worden was the command module pilot. Call sign for the command module was *Endeavor* and for the landing craft, *Falcon*.

Buckbee welcomes Apollo 15 Commander Dave Scott to Huntsville

The Huntsville lunar roving vehicle call sign, *LRV-1*, gave the astronauts legs while exploring the moon. Often referred to as the moon buggy, it was a sporty driving vehicle for the lunar surface. The astronauts considered the vehicle a spacecraft. That's why it had a hand controller, instead of a steering wheel, mounted between the seats and accessible to either astronaut to control. Built by The Boeing Company and General Motors, it was delivered in 17 months, the shortest delivery time of any manned spaceflight vehicle.

The rover weighed 462 Earth pounds but on the moon only 77 pounds because of the 1/6 gravity field. It was about 6 feet wide, 10 feet long and powered by electric motors mounted on each wheel. It could be steered by the rear or front wheels.

The ability of Scott and Irwin to explore the lunar terrain was greatly enhanced by the use of LRV-1. After they unfolded the machine from the landing craft, they began history's first drive on the moon. It was obvious they were having a ball: *"Here we go!...Remarkable machine!"* and, *"This is the only way to travel!"* were among many transmissions. Before the second excursion, Commander Scott remarked, referring to a steering problem they encountered on the first traverse said, *"I bet you had those Marshall guys (Huntsville engineers) come here last night and fix it."* Both astronauts noted seatbelts were necessary while driving the rover on the moon. The ride became bumpy at times, requiring the astronauts to hold on and tighten up their seatbelts. They traveled over 17 miles during the stay on the moon.

Upon the return of the Apollo 15 crew, NASA determined they had engaged in improprieties involving specially stamped envelopes they had taken to the moon—and later sold. They were removed from flight status.

As a public affairs officer, I remember we had numerous inquiries about the moonwalkers selling stamped envelopes back on planet Earth. I thought NASA overreacted, but the media made such a fuss that NASA leadership felt forced to take action. Few people knew that the average salary for a space pioneer was about $20,000 in those days. Today, astronauts and former NASA workers sell NASA memorabilia at auctions and over eBay and never look back.

Scott eventually was assigned to the Apollo-Soyuz program and later became the director of NASA's Dryden Flight Research Center. In 1977, he resigned from NASA and entered private industry. Jim Irwin resigned from NASA in 1972 and established the High Flight Foundation, a non-profit evangelical organization which arranged worldwide climbing expeditions and retreats led by Irwin. An avid runner, mountain biker and accomplished climber, he visited Huntsville and the U.S. Space & Rocket Center on several occasions. During one of those visits, he conquered Monte Sano Mountain on his bike before breakfast. Irwin passed away in 1991.

Apollo 15 moonwalker James Irwin, a frequent visitor to Huntsville, is shown here with the museum's A-12 spy plane, predecessor to the SR-71 that he flew prior to becoming an astronaut

Apollo 16 was the second mission to the moon to have a lunar roving vehicle onboard. The commander was John Young, one of the most experienced space fliers in the world. Charlie Duke, a space flight rookie, was the lunar module pilot. T. K. Mattingly, another rookie, was the command module pilot. Call sign for the command module was *Casper* and for the lunar module *Orion*.

Apollo 16 Commander John Young on the moon , who later sent a message wishing Baker a Happy 25th birthday and thanking her for preparing the way

Young, a legend in the brotherhood, has flown six space missions; two aboard the two-man Gemini, an Apollo Saturn lunar orbit, one moon landing, was commander of the first Space Shuttle mission with Bob Crippen and was the commander on the first Spacelab mission.

A Floridian, Young didn't endear himself with the citrus industry during the Apollo 16 mission, when he told Duke on the moon that Tang, the drink of astronauts which was laced with potassium to increase their endurance, was causing him to have gas. *"I don't want to ever drink another glass of orange juice...I haven't eaten this much citrus fruit in 20 years. But I'll tell you one thing; in another 12 days, I ain't never eating any more. And if they offer me potassium with my breakfast, I'm going to throw up. I like an occasional orange. I really do. But I'll be damned if I'm going to be buried in oranges,"* said Young. They discussed the subject for some time while on the moon until mission control told them they were talking to the world over an open microphone. It turned out the microphone was stuck. Young apologized, *"Sorry about that...It's terrible being on a hot mike sometimes."* I have never met an astronaut who loved Tang.

Young and Duke gave the rover a real workout on their mission. They performed some endurance tests, setting a speed record of 11 mph while executing extreme turns and skids. Duke provided the commentary while "Barney Oldfield" Young drove. *"He's got two wheels off the ground. It's a big rooster tail out of all four wheels and as he turns, he skids; the back end breaks loose just like on snow. Come on back, John... I never saw a driver like this...The Grand Prix driver is at it again. Hey, when he hits the crater it starts bouncing. That's when he gets his rooster tail. Hey, that was a good stop, those wheels just locked."*

Lunar Roving Vehicle gets work out by Apollo 16 Commander John Young

Even at age 74, Young was listed in 2004 as an active astronaut in the NASA astronaut corps. He talked about the Saturn V, calling it *"a high risk operation, riding a Saturn V to the moon powered by hydrogen. Up to that time the only thing I knew that had hydrogen in it was the Hindenberg. That didn't turn out too good."* Young described staging of the big Saturn, "like a great train wreck."

Young did retire later in 2004, after 42 years of being an active space flier. He was the last of the moonwalkers eligible to fly. He still has strong opinions and speaks out about crew safety, future space travel and pushing the human spaceflight envelope. He occasionally writes terse memos about *"weak-knee bureaucrats who get in the way of making things happen."* As a special assistant to the director of NASA's Johnson Space Center, he was influential and became the in-house conscience of NASA's human spaceflight program.

Charlie Duke served as CapCom during Apollo 11. When Armstrong radioed to the world, *"Houston, Tranquility Base here. The Eagle has landed,"* Duke responded in his Carolinian accent, *"Roger, Tranquility. We copy you on the ground. You had a bunch of guys about to turn blue down here. We're breathing again."* He was referring to Armstrong nearly running out of fuel before he landed.

Duke resigned from NASA in 1975 and went into private business. He is very active as an inspirational speaker, contributing a great deal of his time visiting and speaking around the world as a Christian lay minister. A frequent visitor to Huntsville and U. S. Space Camp, Duke was a spokesman and test driver of Mercedes' M-class, sport utility vehicle built and unveiled in Alabama in 1998. Duke drove the new high tech vehicle over a special obstacle course at Vance, Alabama, comparing it to driving the lunar rover on the moon.

Duke test-drives the M-Class over an obstacle course at the Mercedes-Benz U.S. International headquarters in Vance, AL.

Charley Duke, Apollo 16 moonwalker and lunar rover driver, is shown here introducing the new Mercedes-Benz M-Class SUV

Apollo 17 had a lot riding on the flight. It was the last mission to the moon for the country. Gene Cernan was the commander and the last human to make a footprint on the lunar surface. Cernan is another astronaut who almost didn't make it to the moon because of an earlier aircraft incident. He crashed a perfectly good helicopter in the Indian River near Kennedy Space Center and lived to fly to the moon.

Harrison "Jack" Schmitt, the first scientist-astronaut to fly in America's space program, accompanied Cernan. His fellow astronauts referred to Schmitt, a rookie and geologist, as Dr. Rock. The scientific community, who had been lobbying Congress to put a scientist on Apollo, had won. There had been plans for at least three more moon landings but with public interest dwindling and budget cuts, NASA cancelled the remaining flights. The brotherhood wasn't happy when Schmitt got the Apollo 17 seat assignment because he knocked Joe Engle, a good stick and rudder man, out of a ride to the moon. The third member of the crew was command module pilot Ron Evans, who was referred to as Captain America by his crewmates. Call signs were *Challenger* for the command ship and *America* for the lander.

Many space workers lost their jobs when Apollo 17 lifted off. Pink slips were being delivered throughout the Apollo contractor community as the program came to an end. Cernan and his crew spent as much time motivating workers to make Apollo 17 the best, as they did training for the mission.

The last Apollo to the moon was the program's most ambitious. Cernan and Schmitt spent more time on the moon's surface, 22 hours and 5 minutes; set a speed record of 11.1 mph in the rover; covered the most distance in three treks, 19.9 miles; and collected and returned more rocks, 248 pounds, than any other mission. During one of their treks, the explorers came across the most exciting find of the moon missions.

Cernan: *There is orange soil.*

Schmitt: *Well, don't move it until I see it.*

Cernan: *I've stirred it up with my feet.*

Schmitt: *Hey, it is. I can see it from here. It's orange. Wait a minute; let me put my visor up. It's still orange.*

Scientists later determined the soil to be tiny spheres of colored glass that erupted from the depths of the moon in some distant age.

After completing their final exploration of the surface, Cernan said, *"As I take these last steps from the surface for some time to come, I would just like to record that America's challenge of today has forged man's destiny of tomorrow. And as we leave the moon and Taurus-Littrow, we leave as we came and, God willing, we shall return, with peace and hope for all mankind."*

Cernan retired from NASA and the Navy in 1976. A handsome, gifted speaker, Cernan can still charm an audience with his charisma and message, "I was the last man to walk on the moon." He offered moving remarks at a memorial service for Louise and Alan Shepard in 1998, where he represented the astronaut corps and personally paid tribute, *"to that classy lady married to the Navy fighter pilot and America's greatest astronaut and moonwalker, Al Shepard, who I proudly served as backup on Apollo 14."*

Schmitt resigned from NASA in 1975 and was elected to the U.S. Senate from New Mexico. While in Washington, he picked up the nickname "Moon Rock Jack." He served a six-year term, devoting a lot of time to raising the level of awareness among his fellow senators of the importance of science and technology to future generations. Today, he remains active with the aerospace community, several universities and research organizations. He is chairman of the Albuquerque-based Interlune-Intermars Initiative and is a leading advocate for commercializing the moon. In an October 2004 by-line article in *Popular Mechanics*, Schmitt said, *"I believe that helium-3 could be the resource that makes the settlement of our moon both feasible and desirable."*

In a box canyon in the Taurus-Littrow region of the moon, deeper than the Grand Canyon, resides *Challenger*, the craft that brought the last explorers from Earth to the moon. Attached to the craft's landing gear is a plaque left by Cernan and Schmitt which reads, "Here man completed his first explorations of the Moon, December 1972, A.D. May the spirit of peace in which we came be reflected in the lives of all mankind."

Apollo 17 crew and rover with mountains in the background

Reenactment of the first moon landing on the
moon crater, Space & Rocket Center, 1989

I asked Gene Cernan to offer what it was like to ride the Saturn V. Here's Cernan's account:

As I look back on Apollo, I recall a special memory of the people of Huntsville, Alabama, and Wernher von Braun's rocket team. As a moonwalker and the last man to walk on the moon, I owe you a debt of gratitude. You provided me not one, but two great rides on Saturn V; one on Apollo 10 around the moon and the other on Apollo 17 when we landed in a mountainous valley. You also furnished me a rental car for that excursion on the moon. It turned out to be the ultimate driving experience, the lunar roving vehicle.

When Wernher von Braun told my astronaut colleagues and me that he was going to provide us a super rocket machine that would take us to the moon, I had no idea what he envisioned. To be one of the privileged humans on this Earth to ride a Saturn V TWICE is very special. I described this experience in my book, "The Last Man on the Moon," published by St. Martin's Press.

On Apollo 10, I was atop the Saturn V with my crew members, Tom Stafford and John Young, when we lifted off from the Cape. We experienced a deep, muted growl, feeling as much as hearing it as the vibration rolled up the steeple and shook the insides of the spacecraft. There was no turning back at that point. The giant rocket had come alive and its horrendous might was absolutely terrifying. I was pressed hard into my couch as the seconds ticked away. I held my breath since I wasn't accustomed to leaving the launch pad on schedule. I wondered how anything could still be working in the pit of roaring fire.

The enormous clamps securing us to the pad snapped back and our Saturn stirred, the giant nozzles vomiting fire and swiveling to keep the nose pointed straight upward. The rocket gained balance and after a heartbeat or two, it rose from the pad. Up we went, trailing a thundering jackhammer of incendiary orange and white flames more brilliant than a welder's torch.

The Saturn tore through a coverlet of high clouds that partly obscured the sun, which glowed no brighter than our rocket, and rolled lazily onto an eastern course that would lead us to a parking orbit around Earth. "What a ride," I reported to the world as the g forces pressed me against my canvas couch. In a minute and a half we soared a dozen miles and were pulling 4½ g's. The thrill was marred by the onset of a so-called pogo motion, which shook us up and down as if the gods were mixing martinis. We had expected it. No worries and the first stage rocket engines pushed that incredible huge load, growing lighter by the millisecond as the fuel was devoured, towards a maximum speed of 4,500 mph. After only a few minutes, we were already 46 miles high, ready for the first stage engines to run dry of fuel and stop, letting the second stage kick in.

We anticipated the same sort of staging jolts similar to the early Gemini flights, but this shutdown was unusually violent and we were thrown back and forth against our straps as hard as if we were hitting a wall. The trailing fireball swallowed us whole. John Young likened the view to "a great train wreck."

Then, the second stage fired and we were snapped back into our seats again, and we blasted through the fireball in a blink. Things happened so fast on a Saturn. The pogo stayed with us, worse that ever, as another million pounds of liquid hydrogen and liquid oxygen fuel burned hot and hard for seven minutes. We accelerated with breathtaking speed. We weren't done yet: The spacecraft was talking to us, and we didn't like what it had to say. Low moans and a creaking groan indicated that metal was aching and straining somewhere in the back of us and the pogo wobbling indicated that big trouble was brewing down below. What the hell was going on back there? I wanted a rearview mirror that would let me watch all the excitement.

As we were still flying atop something as big as a skyscraper, we couldn't actually see what was happening in a place equivalent to 20 stories below us. Then the escape tower, which we no longer needed, blew away with a loud thunderclap. It was another Saturn surprise, even though I was expecting it. The detonation was so sudden and fierce, I wondered for a bleak second if it would tear our spacecraft off the rocket.

The second stage firing ended and left us whistling along at 15,540 miles per hour, not quite in Earth orbit. Despite the rocking and rolling, it was exhilarating to be back in space. As John Young said, *"Just like old times."*

We got another jolt of acceleration from the third stage rocket engine as it kicked in for three minutes, long enough to park us in orbit 116 miles high at a speed of 17,540 mph. Suddenly, that old Saturn was riding like a Cadillac. The pogo vibration vanished when we dropped off the second stage and coasted into a weightless Earth orbit. There is no euphoria at the phenomena of zero gravity for we had all experienced it before. We were much more concerned whether the violent liftoff had caused damage that might wreck our mission.

We restarted the third stage engine and rode that bucking creature on a path to the moon. We burst through to 20,000 mph and finally up to 24,300 mph as Apollo 10 creaked and groaned like an old house in winter. Stay with us now, baby. Come on, burn! We guessed right; the third stage shutdown right on schedule. The bouncing stopped and silence reigned as we slid into a free coast to the moon. I was relieved to find myself still in one piece. Wernher's people built a hell of a machine.

On Apollo 17, I rode in the left seat. That's the seat for the commander. With me was fellow moonwalker Jack Schmitt and command module pilot Ron Evans. This was the only night launch of the Apollo moon landing program. We lifted-off just after midnight. The hold down arms released and Saturn stirred, balanced on a dazzling fireball that seemed to grow to the size of an atomic bomb. As a show-stopping spectacular, nothing in the entire space program compared to our Saturn night launch. *"The clock has started,"* I told launch control. *"Thrust is good on all five engines,"* CapCom responded. Music to my ears, we're on our way. The vibration wiggled up to the tower and shook us as the big rocket came alive, an infernal growing and growling and seemingly out of control. But it wasn't. I held all of its tremendous power in my hands. Should its internal guidance fail, the awakened Saturn would now respond to me, go where I wanted and do exactly what I told her. I had the power to steer her into the heavens or close her down. Prior to the bolts blowing far below, someone else had been making the decisions, but from here on, I called the shots for the shaking, quaking monster and I happily endured every jolt, for this was the pay-off. For good, bad or worse, the next 13 days I was responsible for whatever happened.

Up we went, crashing through a feathery layer of mist, a beautiful shot in the dark. Into a roll at 12 seconds, the g-forces pushing me down while the rocket pushed me up, soul-searching a lifetime as we swept away from the pad, trailing a searing tail a half-mile long and so bright it lit up the night sky from North Carolina to Cuba.

At liftoff the reflected light from our five Saturn engines bounced off the clouds to paint our instrument panel a violent red. All systems were working perfectly and Ron yelled, *"Whoopee!"* Jack cried out, *"We're going up, man, oh man!"*

At 2 minutes and 40 seconds, the first stage burned out and a fireball the likes of which I've never seen gobbled us up, a maelstrom of flame that toyed with my certain knowledge that we were not burning up. We're not. This was normal. The straps holding us to our couch strained as we were slumping forward, backward and side-to-side, again and again. When the second stage fired, people on the ground saw the separation as the explosion of a minor blue star and we catapulted right through the ominous fireball, stacking on ever more speed.

The escape tower separates, barely noticeable in daylight, tearing away in a blinding burst of light akin to the birth of a comet and yanked off the shroud. Soon, Jack would start to jabber like a kid at the circus. At four and a half minutes I reported, *"Let us tell you, this night launch is something to behold."* We rushed out into the star-dotted heavens and dumped the second stage at about nine and a half minutes, then slipped into Earth orbit on the third stage only a dozen minutes after lift-off as easily as falling into a favorite easy chair.

Houston signaled "go," to re-ignite the third stage, which only eased us against our seats. This burn lasted a bit longer than normal, to make up for the time lost during delay on the launch pad. When it was done, we were out of Earth orbit and on our way. Next stop, the moon.

The other transportation system developed by Huntsville's Marshall Space Flight Center was von Braun's favorite car, the lunar rover. When we reached the moon, we unloaded the rover, which was carried outside the Challenger, our lunar lander, like a piano strapped to a truck. With lanyards, cables and hinges we lowered it, a puzzle of folded wheels, armrests, seats, consoles, footrests, fenders, battery covers, antennas, cameras and so many other parts I felt like I was putting together a Christmas bike. Co-moonwalker Jack "Dr. Rock" Schmitt made the highly technical observation, *"It's safe to say this surface was not formed yesterday."* And to think I hauled him all the way to the moon for him to tell me this.

After assembling the rover, I bounced up sideways into the driver's seat as a teenager might hop into an open Jeep, turned on the batteries, tried the steering, checked the forward and reverse controls and goosed it. This was the moment of truth because if it didn't work, we would be walking. Electric motors on each wheel hummed and I fed it more power and scooted away for a test drive around the lunar lander. *"Hallelujah, Houston! Challenger's baby is on the roll!"* A moon-mobile with wire wheels and no top is pretty cool.

Moon vehicles of all sizes and descriptions were brought to Huntsville for von Braun to test drive by space corporations; Grumman, Bendix, General Motors, Boeing and Teledyne Brown.

"Ok, who bent the fender on Rover 3?"

We spent nearly three days exploring the moon's surface and gathering more rocks than any other crew. I climbed on the rover one last time and parked it at a special place one mile away from the lunar lander. It now would become a mobile TV truck broadcasting our takeoff the next day. As I dismounted, I took a moment to kneel and with a single finger, scratched my daughter Tracy's initials, T. D. C., in the lunar dust, knowing those three letters would remain there undisturbed for more years than anyone could imagine.

On my way back to Challenger, I noticed the American flag that I had deployed which had been first carried to the moon and back by Neil Armstrong and Buzz Aldrin on Apollo 11. It was my honor to be the last American to salute the flag on the moon.

There was a sense of eternity about Apollo. Sir Isaac Newton once said, *"If I have been able to see further than others, it was because I stood on the shoulders of giants."* Every man and woman who put in long hours to get us to the moon now stood with me beside the lunar lander, American flag and the lunar rover. These were the giants upon whose shoulders I stood as I reached for the stars. As the last man on the moon, mine would be man's last footsteps on that surface, for too many years to come. To get here, I had ridden man's greatest space machine and driven the ultimate driving machine, both provided by my friend, Wernher von Braun and his team of space transportation experts. Thank you, Huntsville, for great rides.

Apollo 17 moon buggy stripped down for high speed test: Driver Moonwalker Gene Cernan

Last man on the moon-Apollo 17 Commander Gene Cernan

Lift-off of Apollo 17 moon lander Challenger, last mission to moon. Just before lift-off Cernan said to Schmitt, "Let's get this mother out of here."

CHAPTER 20
Landing on Mars in 1982
by Wernher von Braun

On a flight to California to appear on the *Tom Snyder Show*, Shepard and I were enjoying the scenery over the Grand Canyon. Shepard looked down and said, *"It looks like Mars."* I asked how a trip to Mars would compare to his trip to the moon and would he want to go. He said, *"You bet I would go and of course, I would be the commander. The crew would be bigger, probably six or more, and it would be an international crew made up of U.S., British, Japanese, Canadian, Russian and Chinese space flyers."*

Waxing upon his topic, Shepard continued, *"The mission might take two years. Training would be tough: It would require a lot of preparation because of the long duration flight. I think they would launch us into Earth's orbit where we would dock with a nuclear-powered Mars ship for the trip out. Once we launched out of orbit, it would be a long, boring journey. Once we got into the vicinity of Marsville, things would begin to get interesting. We would probably split the crew, with half going down to the surface and the others remaining in orbit. I would, of course, fly the ship down and land and be the first human to step on Mars. I would plant the U.S. flag and invite my fellow international crew members to do the same."* That was Shepard's short version of a Mars mission.

There had been a plan to go to Mars. Just 20 days after Neil Armstrong and Buzz Aldrin walked on the moon, Wernher von Braun stood in front of the Senate's Space Task Group in Washington, D.C. and presented a plan to send an expedition. I was delighted to read von Braun's 1969 paper. It was well conceived and outlined steps over the next dozen years for a landing on Mars. The following are highlights of von Braun's presentation:

With the recent accomplishment of the manned lunar landing, the next frontier is manned exploration of the planets. Perhaps the most significant scientific question is their possibility of extraterrestrial life in the outer solar system. Manned planetary flight might provide the opportunity to resolve this universal question, thus capturing international interest and cooperation.

What I am presenting today describes a method of landing man on the planet of Mars in 1982. The scientific goals of the mission are described and the key decision stages are identified. The unmanned planetary missions are critical to the final designs selected.

Although the undertaking of this mission will be a great national challenge, it represents no greater challenge than the commitment made in 1961 to land man on the moon.

Landing man on Mars is a program to be undertaken over the next two decades. The systems and experience resulting from the Apollo program and missions proposed for the 1970s provide the technical and programmatic foundation for this undertaking. A 1982 manned Mars landing is a logical focus of the program for the next decade.

The mission begins with the boost of the planetary vehicle. Following assembly of the vehicle in Earth orbit, the earth departure phase of the mission is initiated. The Mars vehicle then begins a 270-day journey to Mars. This is by no means an idle phase of the mission. In addition to observations of Mars, many other experiments and measurements will be made on the Earth-to–Mars and Mars-to-Earth legs of the trip that are of prime scientific importance. The spacecraft represents a manned laboratory in space, free of the disturbing influence of the Earth. The fact that there will be two observation points, Earth and the spacecraft, permits several possible experiments regarding the temporal and spatial features of the interplanetary environment. In addition, the spacecraft can be used to supplement and extend numerous observations conducted from Earth orbital space stations, particularly in the field of astronomy. It is possible, for example, that as yet unidentified comets might be observed for the first time.

Upon arrival at Mars, the spaceship engine fires slowing into orbit in the same fashion that the Apollo moon ship was placed into lunar orbit. The ship remains in Mars orbit for about 80 days during which time the Mars surface sample return probes and the Mars excursion module are deployed and the surface exploration takes place. At the end of the Mars capture period, the spacecraft is boosted out of Mars orbit. The return leg of the trip lasts about 290 days during which many experiments and observations are again conducted. A unique feature of the homeward trip is a close encounter with Venus about 1,230 days after departing Mars. Probes will be deployed at Venus during this phase, in addition to the radar mapping measurements that will be made.

The approximate two-year journey ends with the return to Earth orbit and following medical examinations the crew will return to Earth via the Space Shuttle.

The Mars landing mission can be accomplished with a single planetary vehicle assembled in Earth orbit. There are, however, advantages in deploying two ships on the mission because of the long duration. One obvious advantage is in crew safety, each spacecraft being designed to accept the crew of the sister ship in the event of a major failure. This approach also allows more exploration equipment to be carried on the expedition and enhances the probability of achieving mission objectives.

In the current concept, each vehicle assembled in Earth orbit consists of three nuclear propulsion modules, side-by-side, with the planetary spacecraft docked to their center module. Each spacecraft is nominally capable of sustaining a crew of six people for two years, or a crew of twelve for an extended period, in case of an in-space emergency.

The Earth orbit departure maneuver is initiated with the firing of the outer two propulsion modules.

The two planetary vehicles are assembled in a circular Earth orbit. Each vehicle consists of a spacecraft, two nuclear shuttle vehicles for Earth departure propulsion, and one nuclear shuttle for the remaining propulsion requirements through the Mars mission.

Following assembly and checkout in Earth orbit, each of the planetary ships is accelerated by the two outer nuclear shuttles to trans-Mars injection velocity.

The two outer nuclear shuttles are then shutdown, separated from the planetary vehicle, oriented for retrofire, and then retrofitted to place them on a highly elliptic path returning to the original assembly orbit altitude. After a coast of several days, the nuclear shuttles arrive at the original assembly orbit altitude and are retro fired again to place them into a circular orbit. The nuclear shuttles rendezvous with a space station to be checked out and refueled for further utilization.

The nuclear shuttles, which return to Earth orbit, will be available for transfer of fuel and supplies to geosynchronous orbit or to lunar orbit.

The forward compartment of the spacecraft is an unpressurized area housing the Mars excursion module (MEM), an airlock to provide for pressurized transfer to the MEM and for extra vehicular activities and unmanned probes. Six Mars surface sample return probes and two Venus probes are carried on each spacecraft.

Immediately aft of the airlock is the mission module that provides the crew with a shirt-sleeve environment, living quarters, space vehicle control capability, experiment lab, radiation shelter, etc. The functional areas are distributed on four decks. This compartment is occupied by the crew of six for the entire mission except during the Mars surface activity.

At the aft end of the mission module adjacent to the nuclear shuttle is a biological lab where the Mars surface samples are received and analyzed. The biolab is sterilized prior to Earth departure and remains sealed until initial remote analyses of these samples has been accomplished.

The ability of man to withstand a zero gravity environment for periods of time exceeding a few weeks is still an unknown. The Skylab to be flown in the early 70s will determine man's capabilities in a zero gravity environment for a few months. It will remain, however, for the space station to demonstrate man's capabilities for the longer periods required for the manned Mars landing mission.

The option to provide artificial gravity for the crew during the planetary trip must be kept open until conclusive results of man's abilities are established. If early missions indicate the need for artificial gravity, the two spaceships can be docked end-to-end and rotated in the plane of the longitudinal axis during extended coast periods.

During the outbound coast to Mars of approximately nine months, experimental activities such as solar and planetary observations, solar wind measurements and biological monitoring of the crew, test plants and animals, will be conducted. At the end of the period, final space vehicle checkout for the Mars orbit insertion maneuver is followed by retrofires of the nuclear stage to place the planetary vehicle into an elliptical Mars's orbit. The orbit at Mars is elliptical both to reduce energy requirements for the mission and allow a wider range of planet coverage by optical observations. The first two days are used to select landing sites for the unmanned sample return probes.

On the first manned mission, it may be desirable to obtain samples prior to the actual landing of man and subsequent contamination of the planet. Surface samples can be obtained with sterile unmanned probes deployed from the manned spacecraft. The probe would descend from the orbiting spacecraft, land on the Martian surface, automatically gather a sample and return it to the biological lab in the spacecraft for analysis. If the analysis revealed no significant biological hazards, man can then proceed to the surface.

The three-man landing party from each ship is carried from the orbiting spacecraft to the surface in the excursion module. Except for the effects of the Martian atmosphere, the landing and return to orbit sequence is analogous to the Apollo lunar landing operation utilizing the lunar module. In the case of Apollo 11, the Mars orbiting spacecraft is like the Columbia mother ship and the MEM, excursion module, is comparable to the Eagle or landing craft.

Following final checkout, the MEM is separated from the spacecraft and de-orbited by the retro-firing of a small rocket motor. The MEM is then aerodynamically decelerated as it falls toward the surface. As the MEM approaches the surface, the protective shroud and portions of the heat shield are jettisoned. Jettisoning of the shroud allows use of the ascent stage as an abort vehicle, if required before landing. The descent stage engine then provides terminal braking and hovering just prior to touchdown. The MEM then spends 30-60 days on the surface of Mars.

At the conclusion of surface operations, the ascent stage is fired to initiate the return to Mars orbit and the waiting spacecraft. Propellant tanks are staged or discarded during ascent to save weight. After achieving proper orbit conditions, the MEM will rendezvous and dock with the spacecraft. Following docking, the crew will transfer to the mission module and the MEM will be discarded.

The Apollo-shaped Mars excursion module is designed to carry three men to the surface of Mars and return the crew, the scientific data and samples to the spaceship. It provides living quarters and a lab during the 30-60 day stay on the Mars surface. The MEM consists of descent and ascent stages. The ascent stage houses the three-man crew during entry, descent, landing and ascent. The ascent stage consists of the control center, ascent engine and propellant tanks. The descent stage contains the crew living quarters and lab for use while on Mars, the descent engine, propellant tanks, landing gear and outer heat shield for the aerodynamic entry phase of the descent. A small, one-man rover vehicle is provided in the descent stage for surface mobility. All descent stage equipment is left on the Martian surface. The capability is provided to land a MEM and bring a stranded crew back to the ship orbiting Mars.

Man's first step on Mars will not be less exciting than Neil Armstrong's first step on the moon. The Mars surface activity on the first mission will be similar in many ways to the Apollo 11 moon surface activity. Notable, however, is the much longer stay time (30-60 days per MEM), thus allowing more extensive observations, experimentation and execution of mission scientific objectives. The small rover vehicle allows trips to interesting surface features beyond the immediate landing area. Surface operations included experiments to be performed in the MEM laboratory, as well as the external operations on Mars' surface.

During the planetary surface operations, the men in the orbiting spacecraft continue their experimentation observations, monitor the surface operations, and maintain the necessary space operations.

Using the MEM descent stage as a launch platform, the ascent stage delivers the crew, scientific data and samples back to the orbiting spaceship. The return payload, consisting mainly of samples, data and miscellaneous equipment, weighs approximately 900 pounds.

At the completion of the 80-day period at Mars, the planetary spaceships will begin the return leg of the journey. The nuclear stage is ignited for this propulsive maneuver, boosting each spaceship out of Mars orbit.

With extensive Mars exploration activities behind them, the crew at this point can began a more thorough analysis of the data and samples gathered at Mars and prepare for the next major milestone of the trip—a close encounter with the planet Venus. This return trajectory provides an opportunity for close-proximity observations and experiments at Venus. Two probes are provided on each spacecraft and will be deployed during the Venus passage.

The manned Mars landing mission concludes with the return to Earth orbit, using the last of the propellants in the nuclear stage for the braking maneuver. An optional Earth-returning mode would allow the crew to make direct aerodynamic entry, Apollo style. Until a better assessment can be made of the contamination hazard, the return by man of pathogens that might prove harmful to Earth inhabitants, a more conservative approach has been planned, i.e. the return of the crew to Earth orbit for a quarantine period. Another advantage of the orbit return mode is that the nuclear stage and mission module are available for possible reuse.

Once the spacecraft achieves the desired Earth orbit, it will rendezvous with the waiting space base (space station), where the crew will receive thorough medical examinations before returning to Earth via the Space Shuttle. The returned samples could also be further examined prior to their return to Earth.

Perhaps the single, most consuming scientific question of the space program is: Does extraterrestrial life exist in our solar system? Has life ever existed on Mars? Does it exist now? Are conditions such that some forms of life could exist? Man on Mars will be able to study not only the forms of life indigenous to Mars, but also the behavior of terrestrial life forms transplanted to the Martian environment..

Dr. Thomas O. Paine, the NASA administrator in attendance at that presentation, was asked his and the senators' reaction. *"My recollection is that he (von Braun) was treated with tremendous courtesy and respect, and that they received his presentation very thoughtfully and very seriously,"* responded Paine. *"However, I feel that they shared the judgment of the other end of Pennsylvania Avenue (the White House) that the interests of the American public were elsewhere and that as a politically feasible thing, NASA was dwindling away. You can think of it as a Churchill-like historical occasion that was not going to lead to anything."*

In 2004, I shared von Braun's presentation with Schirra and he was impressed. *"We need to get this message out. Let people know von Braun and his team were working on going to Mars in the '60s. Amazing."* I asked Schirra what he thought of President George W. Bush's vision for space exploration. He commented, *"I like what President Bush is trying to do…He wants to open up the space frontier again, go back to the moon and on to Mars. In the next four, five or ten years, we must get our steps in order—get our young people excited about the project. He didn't say that exactly. He was trying to get the country excited by racing back to the moon. We aren't racing with anyone. I don't believe we have to go to the moon to go to Mars. That's my opinion. I happened to have been educated by von Braun in that regard."*

To GAYLE
with happy memories
Pathfinder

Alan Shepard

Shepard with Buckbee and wife Gayle during
dedication of Space Shuttle Pathfinder in Huntsville

MERCURY-
REDSTONE

Shepard and Buckbee at 30th anniversary
celebration, with actual Mercury Redstone in
Huntsville that flew inches and aborted

Shepard's thoughts on life after being an astronaut were quite different from his peers. He was of the opinion that in the early days, astronauts weren't trained to be anything but fighter pilots and astronauts. He didn't feel he and his colleagues had the training to become heads of corporations or heads of state. Consequently those job offers and suggestions went unheeded. I know the Kennedys— Jack and Bobby—had encouraged him toward politics, probably Johnson, too. He felt this was more of John Glenn's character, as he would say, *"One Mercury astronaut in politics is enough."* When Glenn ran for president and lost in the 1980s, he had a major debt to pay and the law prevented him from paying it himself. Consequently, Glenn had to ask his Mercury buddies for help. Their answer was, *"Yes, we'll help you John, under the condition you'll promise to never run for again for president."*

After retiring in 1974 from NASA and the Navy, Shepard made an excellent living from real estate, stocks, and banking. Compared to many of his space flying buddies, he was quite well fixed. It didn't hurt that he was quick to accept freebies and ready to take advantage of a good deal when he saw one. And while he could easily be classified as frugal with his money, it was equally true he became increasingly generous with his time. Shepard worked tirelessly on behalf of the Mercury 7 Foundation as their first president, taking no honorariums or fees for himself for public appearances. He had changed. He wanted to give something back.

He called me one day, out of the blue, asking about Space Camp. He wanted to take a look around and find out more about it. This led to several visits and his growing involvement. *"Buckbee, I like that Space Camp thing you're doing,"* he'd say, *"generating all those little wannabe Shepards all over the world."* He was still the cocky, egotistical and sarcastic Shepard, raised and educated to be a leader in a very dangerous and unforgiving profession. But he also had found a way of sharing himself with the younger generation who could be motivated by his spirit and accomplishments.

During a period of about 10 years, I traveled with Alan Shepard selling Space Camp and the NASA space program in some 25 states and seven foreign countries. It was not uncommon for Shepard to be recognized in an airport and approached by someone just wanting to shake his hand and thank him for his courage and commitment to the nation. You could sense that José was pleased, graciously thanking them in a manner unlike the "icy commander" from days gone by.

I arranged a number of special events called "An Evening with Alan Shepard." Coca-Cola USA, United Technologies Corporation, Macmillan/McGraw-Hill, Brunos Golf Classic, Pebble Beach Golf Club, Rockwell International, Nippon Steel and others rented museums, science centers and unique settings to have a private evening with America's first astronaut. The honorarium was always donated to the Mercury 7 Scholarship Fund or to Space Camp. One such event, sponsored by Romanoff International, was held on August 18, 1991, at the National Air and Space Museum. Romanoff's had been selling Russian caviar for over 150 years. It was a private affair with about 250 of Romanoff's corporate officials and prestigious accounts. The Air and Space Museum had reserved the Space Gallery where Shepard's Freedom 7 spacecraft was displayed. Shepard, who was introduced by the chairman of the board, made a short speech about how honored he was to join them for the evening. Guests were invited to come to the Freedom 7 and have their picture taken with Shepard. I was amazed how quickly the ladies grabbed a glass of wine and pulled their husbands in line. Shepard answered questions, pointing to key parts of the interior of the spacecraft, and posed for the picture. Shepard took the center position between the couple with his arm around the lady. After the picture was shot, I wrote down the names so Shepard could personalize each photo. It was a long evening but Shepard did his job of greeting and signing lots of autographs. Before we left that evening, I informed the event manager that Admiral Shepard would be leaving the building in 10 minutes. As we exited, we heard the announcement, *"Ladies and gentlemen, Admiral Shepard has left the building."* Shepard loved it, shades of Elvis.

One of Shepard's complaints about special events was the inevitable encounter in the head. It seemed people would watch when he went to the bathroom and that would trigger a rush. Shepard would ask, *"Why is it every time I go to the head, all these guys line up to go? It's like Grand Central Station in there. Not only that, I have to shake hands with a bunch of old farts who take forever to urinate and probably haven't washed their hands."*

I tried to explain to the Admiral that many people considered it an honor to share the bathroom with an original M7 astronaut, not to mention, a moonwalker. It would be a highlight for most guys not to mention a great story to tell their grandchildren.

Shepard liked to tell jokes and he didn't spare the senior citizens. One of Shepard's favorite sayings was, *"old age and trickery will overcome youth and ambition every time."* His old people jokes were numerous. I heard some of them so often, I could remember them by the numbers:

1.   " You can tell you are getting old when you eat your breakfast before you go to bed!"
2.   "You know what they call Nuns at night, roaming Catholics."
3.   An old man setting on the park bench crying. His buddies asked him why are you unhappy today. Yesterday you told me you had just married a beautiful young girl and you were never happier. What happen? The other guy says, 'I can't remember where I live!'
4.   "First pilot says, "Windy, isn't it?" Second pilot answers, "No, It's Thursday." Third pilot's response, "So am I. Let's go get a beer."

Shepard stole one story from his back-up crew member Gene Cernan using this opening line before a NASA audience: *"I feel like Elizabeth Taylor's seventh husband, I know what to do but I don't know how to make it interesting."*

Bobby Wilkinson, the Coca Cola bottler in Huntsville, arranged for Shepard and me to visit Don Keogh, president of the Coca-Cola Company based in Atlanta. Wilkinson, something of a legend in Coca-Cola circles, is the guy who told the powers that be that the famous New Coke product was dead on arrival in the marketplace. Shortly afterward, Wilkinson was quoted in *TIME* magazine, *"The Marketing Lesson for the Decade: What Coca-Cola didn't realize was that the old Coke was the property of the American public. The bottlers thought they owned it. The company thought it owned it. But the consumers knew they owned it. And when someone tampered with it, they got upset."*

Huntsville Coca-Cola president Bobby Wilkinson discusses with Buckbee and Shepard how to get the Space Coke can on the Space Shuttle

When we arrived at Coca-Cola headquarters, Shepard was an immediate hit with senior management, particularly Walter Dunn. Dunn was in charge of Prestigious Accounts for the company. Our aim was to convince the Coca-Cola Company to become a sponsor of Space Camp and the Astronaut Hall of Fame. Later, we would refer to that visit as the $1 million meeting. Over the next several years, the Coca-Cola Company gave more than that amount to the Space Camp and Astronaut Hall of Fame programs.

While in Atlanta, Shepard and I were invited to the Delta Air Lines pilot training center at Hartsfield International Airport. It was a visit ripe for a bit of fun and games with the latest flight training simulators. These were highly sophisticated, Boeing 757 simulators. Shepard was in the left seat, the captain's position, and I was in the right. Shepard reminded me of a procedure in flying to sound off when passing the controls to the other guy in the cockpit. We were on final approach for a second run in the simulator and I was flying the airplane. I was obviously too high and too fast on the approach and at the last moment I yelled *"It's your airplane, José!"* We clipped the tower and crashed into the terminal. According to the simulation officer, we all died in the crash. On the next flight, Shepard took over and flew the airplane. On final approach, just before touchdown, I reached over and moved the flap control lever to the "up" position. With no flaps we were coming in very hot and fast and hit the runway hard, bounced a of couple times, then blew two tires. Shepard, the naval aviator, astronaut and moonwalker, didn't take it well with all those Delta pilots watching.

Buckbee and Shepard, meet with senior Coca-Cola, USA executives, Ike Herbert, second from left and Don Keogh, fourth from left, resulting in a partnership that produced over $1 million in contributions from Coca-Cola to Space Camp and the Astronaut Hall of Fame

On another occasion, Shepard and I went to St. Paul/Minneapolis to negotiate a marketing contract with a national travel agency for Space Camp. At the end of the day, we went to dinner. Earlier, Shepard had reminded us of an eclipse of the moon that was going to occur that night. As we were walking to the restaurant, Shepard commented it wasn't going to be possible to see the moon because of the tall buildings. Someone suggested we try to go to the top of one of the buildings and watch the eclipse from there. We tried and there was no access to the roof. As we were about to leave, the ladies in the party stopped to use the facilities. Upon exiting, one of the women commented that she could see the moon from the ladies restroom. That did it. We all crowded into the ladies room and Shepard gave a description of the eclipse that only a moonwalker could give.

Yes, even if it was from a ladies restroom, Shepard always wanted to command any given situation. But there was one time I didn't let him have his way. We were coming back from Maryland, traveling to National Airport and decided to take a shortcut to save time. After several miles I informed Admiral Shepard that in my humble opinion, we were lost. The Admiral informed me he knew exactly where we were and would reach our destination on time. After another 20 minutes of wandering around Maryland, I again informed him I thought we were lost. He stopped the car, got out and began stargazing. *"Yep, I know exactly where we are, there's the North Star."* He let me know his space navigation training was paying off. *"José,"* I said, *"we're not trying to get to the moon, tonight. We're trying to find National Airport."* Following the Admiral's "on course" directions we ended up in a really bad section of Anacostia. I felt like I was in one of those Chevy Chase vacation movies, on a deserted and dark street where they steal the car's hubcaps at stoplights. About that time I spotted a policeman. Despite Shepard's objections, I frantically waved to the officer and he pulled up beside us. *"Can I help you gentlemen?"* Before Shepard could say anything I said, *"Yes, we're lost and we need directions to the National Airport."* The policeman came around to Shepard's side of the car and Shepard greeted him, *"Hello officer, I'm Alan Shepard."* I interrupted and said, *"Don't pay any attention to him, he always says he's Alan Shepard and I'm sure you understand."* The officer looked at Shepard and commented, *"It's obvious you're not Alan Shepard because he wouldn't be lost two miles from National Airport!"* He walked back to his patrol car while pointing to the airport.

Several years later, I was going to Houston for a meeting with Shepard and he said he'd pick me up at the airport. I was looking for someone in a Corvette or muscle car of some sort when some guy in a Chevy blows his horn. Sure enough, it was Shepard in a Chevy Beretta! I threw my bag in the car, climbed in, and gave Shepard a quizzical glance with a slightly derisive comment about a custom, two-door Chevy. I recall a response that went something like, *"Shut-up Buckbee, you're lucky I'm picking you up at all. I'm 30 years more mature now. Besides it's difficult to get up to 100 mph between my home and the office when I only live three blocks away."*

We had a couple hours to kill before going to see Aaron Cohen, director of NASA's Johnson Space Center. (It was Shepard's desire to establish a Space Camp at the Johnson Space Center.) As we were driving, Shepard began reminiscing. The image of the Mercury Seven as a gung-ho, daredevil bunch is pretty accurate. In the old days, all of the guys enjoyed driving fast and pushing cars not to the total limit but certainly close to the edge. We parked in his old space at Johnson and he regaled me with "car stories." He laughed as he remembered the time Schirra bought a Ferrari supposedly owned by some famous Hollywood actor. It was the 1960s when everyone in the brotherhood but Glenn was driving Corvettes, Shelby Cobras or Austin Healys. Glenn owned a Studebaker and later drove a four-cylinder NSU Prinz which, "might do 60 mph wide open on a flat road with a tailwind." Schirra took special care in cleaning his Ferrari so he could arrive in the parking lot of the astronaut office with the most splendid looking sport car ever made. He didn't know Shepard had heard about his purchase and decided to try a "gotcha." Shepard contacted one of his Indianapolis 500 buddies and got a loaner for one day, an actual Indy 500 all-up race car, plus a one-day driving permit from the local sheriff. With everyone looking at Schirra's Ferrari, Shepard roared up in the Indy car, parked, jumped out with briefcase in hand, and headed for the office.

Several of the M7 were driving to San Antonio, Texas for more medical tests by the medical cult. Glenn took off an hour earlier in his four cylinders Prinz that got 50 miles to the gallon. Shepard and Slayton take off in their speedy Corvettes and passed Glenn who had been pulled over by a Texas trooper. They gave Glenn the high sign and thundered on down the road. Shortly thereafter, the non-sports car driver Glenn, hung a sign in the astronaut office that read: "Definition of a Sport Car: A Hedge Against the Male Menopause."

During the same time period, Shepard headed for downtown Houston in one of his new Corvettes with no license tags. He was stopped for speeding. The officer asked for his license and registration. Shepard realized he had no wallet, no licenses, no registration and no tag on the car and so informs the officer. The officer looks at Shepard and says, *"Oh, that's ok Colonel Glenn. I know who you are. Just slow it down a little."* Shepard response, *"I thank you officer and America thanks you."*

With perhaps one exception, every astronaut I ever traveled with drove the car or flew the airplane. I asked Shepard about that once and he said it wasn't that he didn't trust others to drive or fly, he just believed he was trained and conditioned to be a better driver or pilot.

Once when Shepard and I were leaving the U.S. Astronaut Hall of Fame and U.S. Space Camp Florida in different rental cars on our way to the airport, I bet him 20 bucks I could beat him to Orlando International Airport without exceeding the speed limit. *"You're on,"* said a confident Shepard. We departed Space Camp and traveled the Beeline, a toll road, for the 45 minute drive. We were bumper-to-bumper approaching the first toll booth. Shepard stayed right behind me and I led him into the toll booth that was closed. At the last minute, I turned sharply to the left. It was too late for Shepard to turn so he had to stop and back up. By that time, I cleared the booth and was well on my way to the airport. After a few miles, he caught me again and was right on my bumper as we approached the second toll booth. When I stopped to pay the gentleman my fee, I told him I believed the fellow behind me was driving a stolen car and advised the attendant to delay him because the police were on their way. *"By the way, he may even say he's Alan Shepard or something like that, so don't believe him."* Sure enough, from the rearview mirror I could see the attendant was asking to see Shepard's driver's license. Needless to say, I beat Shepard to the airport by several minutes. When he pulled up to the rental car desk he wasn't a happy astronaut. He explained the idiot at the toll booth asked him all kinds of questions, inspected his car, checked his licenses and registration and caused him to lose the bet. I thanked him, took my 20 bucks and he never knew the difference. Gotcha, Shepard!

When Shepard and I were in the Santa Monica area, he would take me to a bar on Ocean Boulevard called Chez Jay's. It was owned by a buddy of his, Jay Fiondella and was one of his favorite watering holes in the '60s. It was famous for its free peanuts. They were everywhere; in bowls on the tables, on the bar, shells all over the floor—even in the rest rooms. You couldn't turn around without touching a peanut. It was the hangout for Tinsel Town's brightest stars like Frank Sinatra and the Rat Pack, and a stopover for JFK's staffers on the way to Peter Lawford's house, known as the California White House.

Even its owner described it as "just a hole in the wall," but the Hollywood set knew they could eat, drink and carouse, relatively safe from the fans, the gapers and the cameras. Chez Jay's and its 10 dimly-lit tables, was just down the road from Hollywood, but a world away from the plush Polo Lounge, the hyped Spago's and the other watering holes of the Hollywood hills. Here, everyone from Marlon Brando to Linda Ronstadt, from Michael Caine to Helen Reddy, from Richard Burton to Warren Beatty, came to savor the peanuts and privacy. Henry Kissinger even discussed affairs of state at Chez Jay's.

"I just don't know why this little place attracts the people it does," shrugged Jay. Part of the secret is Jay himself. Hollywood loves anyone who is larger than life, and Jay appears in movies and TV, flies hot-air balloons and dives for buried treasure. If anything, Jay largely attributes his success to peanuts. They brought him health, wealth and best of all - publicity. Shepard liked the place because when he went there the word spread like wildfire and the stars would show up to trade stories with America's first astronaut.

In 1996, Alan and I sat with Jay at table 10, located in the rear, and Jay confided, *"In a safe deposit box in my bank, I have the world's only astro-nut. Alan came here a lot before his flight to the moon. He pocketed one of the peanuts and later stowed it away in Apollo 14, inside a 35mm film can."* Jay continued, *"After the flight, Alan called up and said, 'Fiondella, I've got the peanut I took to the moon for you.' He brought it down here and signed an affidavit saying where it had been."*

"Sometimes, I supply peanuts for use in movies and I tell them they are the cousins of the one that went to the moon," he grinned.

Anecdotes tumbled from Jay, gems that sparkled like his best champagne. *"Julie Andrews and her husband, Blake Edwards, often came here to eat. One night, I introduced them to Alan. Julie asked what gave him the biggest thrill on the moon, and Alan held his thumb and forefinger about an inch apart and said, 'When I did that, and could see the Earth between my fingers. That was my greatest thrill.'"*

On our last visit to Chez Jay's, we drifted to the subject of Gus Grissom and the Liberty Bell. Jay, a serious diver who loved to explore wrecks, talked about organizing an expedition to locate, recover and restore Gus's spacecraft. The only things he needed were money and sponsors. Shepard really got excited about the idea and wanted to help. We discussed sponsors and made a tentative list to pursue.

We left Chez Jay's around midnight. We promised to stay in touch. Jay hugged Shepard and said, *"I'm going to find Gus's capsule for you guys."* A few years later, Gus's capsule was recovered by Curt Newport, restored and put on national tour.

International destinations were also part of our travel plans. One trip took Shepard and me to Japan to finalize a contract for a Space Camp with Nippon Steel. Although Shepard came along too late to personally participate in WWII, he was trained by fighter pilots who engaged the Japanese in serious combat. Keeping that in mind, Shepard's humor might have occasionally been border line, as he often referred to the Japanese as, "harbor bombers." I had to remind him that the Japanese did not enjoy our sense of humor.

On one of the many bullet-train rides Shepard proclaimed that the Japanese rail industry stole our technology and said so loud enough for several Japanese seated nearby to hear. They sat quietly and stared straight ahead as Admiral Shepard continued to expound. As we entered the outskirts of Nagasaki, our tour guide began describing the significant landmarks and sharing the history of the city.

After listening for several minutes, Shepard said, *"well it looks like a relatively new city to me,"* reminding those within earshot that this was a target for one of our atomic bombs during WWII.

During press conferences when Shepard was the star attraction, he often opened with one of his favorite American jokes, which would be greeted with total silence. He would ask a question like, *"Why did kamikaze pilots wear helmets?"* It was all I could do to get him to refrain from telling his *old man* and *short-people* jokes. The Japanese press often asked Shepard questions that reminded him that he was *second* to the Russians to go into space and the *fifth* man to walk on the moon. I could see Shepard cringe.

But even Shepard was moved by one of the ceremonial highlights of our visit. Shepard and company were invited to visit a famous geisha house. It was a custom for the geisha to parade through the streets on the way to the geisha house. To protect her from the sun, a person was required to walk behind her, holding a large ceremonial umbrella: It was an honor accepted by prime ministers and heads of state. Astronaut and moonwalker Shepard was asked to participate in the ceremonial trek. Once we arrived, a geisha and her court performed a magnificent program, followed by a traditional Japanese meal. The beautiful geisha was dressed in a multi-colored garment and her face was painted white, which contrasted her dark hair and colorful costume. It truly was a memorable, cultural experience.

Playing golf with Shepard and the Japanese was, to say the least, interesting. Since he was the guest and a true American hero, he was given the privilege of first off the tee, which he graciously accepted. After each of his shots, his Japanese counterparts would use the same club and basically copy his approach to the shot. This went on for several holes before Shepard, exasperated, said, *"...I bet if I hit the damn ball into the water, they'd do the same!"* Shepard was a methodical golfer who took his time, while on the other hand the Japanese were usually nervous, fidgety and chatty—a combination ruinous to Shepard's concentration. After a demonstration of the six-iron shot he made famous on the moon, Shepard's moon golf clinic, we retired to the 19th hole.

Back in the states, the M7 astronauts were in Orlando for one of Henri Landwirth's Give Kids the World shindigs held at the Peabody Hotel. As our group was leaving, Glenn was recognized by the security guard coming off the escalator. Shepard and I were several steps behind and saw the two shaking hands. When Shepard got there, the security guard said, *"Do you know who that was?"* pointing to Glenn. Shepard said, *"No, who was it?"* The guard said with great enthusiasm, *"That was John Glenn who went into space when I was in grade school. He was our first astronaut to orbit in space and I got to shake his hand!"* Shepard asked, *"Do you know who I am?"* The guard answered, *"No, I don't think so,"* Shepard said, *"I'm Alan Shepard."* Nonplussed, the guard responded, *"Oh, nice to meet you,"* not recognizing America's first astronaut. Shepard took it well, telling Glenn the story and they both laughed with Shepard adding, *"I never have been fond of security guards!"*

The next evening a group of us were at the Peabody Hotel bar discussing dinner plans. Several suggestions were made and one restaurant seemed to be the favorite of the group. Shepard said, *"Oh, that's too far away."* I assured him it was 15 minutes, maximum. *"That's just too far to go for dinner,"* he maintained. Incredulous, I retorted, *"And you're the guy who took a 240,000-mile trip to walk on the moon and you think a fifteen minute drive to dinner is too far?"* We enjoyed a nice dinner that evening with the admiral at the hotel.

The Years of Wernher von Braun
A special visual tribute

As I review 1950-1970—the Huntsville years of von Braun—I am awe struck by the astonishing accomplishments of this man in such a short period of time.

Working for von Braun was like working for Thomas Edison. Von Braun was an inventor of rockets. I witnessed the incredible birth of the moon rocket in his laboratory and I had the privilege of watching him extend the human presence to the moon and beyond. He did it in his lifetime. He did it his way and he let all of us watch. He truly was a legend in his own time.

NASA historian Eugene Emme said, "Cruel fate denied Wernher von Braun the chance to buy his ticket as a passenger bound for an excursion in space—his boyhood dream and lifetime goal. Because of Wernher von Braun, however, almost everyone has been brought to the realization that we have been passengers on a spaceship all along— Spaceship Earth. Posterity will not forget him..."

Von Braun was once asked what does it take to travel to the moon? "The will to do it," he replied. Dr Ernst Stuhlinger, von Braun's long time colleague and chief scientist of the rocket team said it best for all of us in describing his boss, "For those who were privileged to work for him, his 'will to do it' was probably his most outstanding feature, but he had many more: his clarity in combining visionary dreaming with realistic planning, his capability to see and to organize a project in its totality, his superior judgement in all engineering questions, his brilliant leadership, his capability to instill enthusiasm in others, and simply his exuberant joy of life which made it pleasant for his co-workers and his friends to be near him."

I'm pleased to share with you this special tribute to Wernher von Braun that represents the significant achievements and accomplishments that occurred from 1950 to 1970 in Huntsville, Alabama when von Braun led America in the conquest of space. This special visual presentation was researched and produced through the courtesy of the University of Alabama in Huntsville on the occasion of the dedication of the Wernher von Braun Research Hall on May 13, 2000. The tribute to Wernher von Braun and the rocket team can be viewed by visitors during business hours at the von Braun Research Hall on the campus of the University of Alabama in Hunstville.

1950

Bumper rockets are launched

Members of the German rocket team

The rocket team moves to Huntsville, with the first contingent arriving on April 1. To overcome any unease or lingering bad feelings after World War II, members of the German rocket team are encouraged to become active in community organizations.

"Within 90 days of our arrival in July 1950, my wife and I were licking stamps and envelopes for the community concert association membership drive."

— Walt Wiesman

Upon Dr. Wernher von Braun's arrival, *The Huntsville Times* publishes a front page article about the scientist, along with a photo of von Braun, his wife and their youngest daughter, Margrit. Von Braun takes advantage of the opportunity to promote not Army missile development, but something else dear to his heart:

Dr. von Braun Says Rocket Flights Possible to Moon
Foremost World Authority Explains That Dollars Are Main Hurdle to Overcome, As Technicians Know Exact Trajectory, Flight Path Of Ship To Be Used

By Bob Axelson

Rocket flights to the interplanetary world are not figments of the imagination, but will be made during the present generation, Dr. Wernher von Braun, project director for the guided missile center at Redstone Arsenal, asserted yesterday ...

"The rocket to the moon is in a similar stage of development at the present time as the trans-Atlantic flight by airplane was in 1912," he explained.

"We scientists know exactly what such a rocket would look like, its size, trajectory, flight path, power plant needed and other requirements. The only thing it takes is development," the 38-year-old scientist declared.

— *The Huntsville Times*, May 14, 1950

Hermes rocket prepared for launch

The Bumper Round 8, the combination of V2 and Corporal missile systems, becomes the first Army rocket launched at Cape Canaveral.

The University of Alabama's Huntsville center opens its first classes on January 6, in rooms at West Huntsville High School. There are 137 students. Ten classes of seven basic freshman subjects are taught by five part-time faculty. By the Spring quarter, enrollment has risen to 234. The average age of the students is about 30 and many students use their G.I. Bill benefits to help pay for tuition.

In September, George Campbell is named the center's first permanent director, and the center's first full-time faculty member, John McCormick, is hired to teach English.

Engineering area

First launch from Cape Canaveral, Florida

1951

Honest John rocket

Corporal rocket

Josiah Gorgas Laboratory

First Redstone launch

The rocket team performs design studies and engineering activities for a new ballistic missile system, which would become the Redstone.

Test facilities at Redstone Arsenal are designed, including the static test stand, later designated a national historic site. Using the same shovel he used to break ground for the Redstone Ordnance Plant in 1941, Col. Carroll Hudson, Redstone Arsenal's commanding officer, breaks ground for the $1.5 million Josiah Gorgas Laboratory.

The British Interplanetary Society hosts its second International Congress on Astronautics in London. The official subject: An Earth-satellite vehicle.

> The delegates attacked nearly every angle of designing, launching, supplying and utilizing satellites, and none had given the matter closer study than Dr. Wernher von Braun ... now hard at work for the U.S. Army in Huntsville, Ala. ...
>
> In considerable detail, von Braun sketched out a full-dress flight to Mars. It could be done, he wrote, by using two satellite stations as intermediate refueling and supply bases.
>
> The fuel required — 5,356,600 tons — is a lot of fuel, but he pointed out that one-tenth as much was burned up during the Berlin airlift "just because of a little misunderstanding among diplomats." He hoped that when mankind enters the cosmic age, "wars will be a thing of the past ... and people will be ready to foot the fuel bill for a voyage to our neighbors in space."
>
> — *TIME* magazine, September 17, 1951
> "Astronaut von Braun, to Mars and back in 969 days"

In October, the Redstone Arsenal Institute of Graduate Studies is established in Huntsville on a contractual basis with The University of Alabama. The goal is to promote the effectiveness of rocket and guided missile programs by improving technical training through refresher courses, seminars and graduate studies.

Honest John rocket

Army personnel secure Nike missile

1952

V-1 on display at Redstone Arsenal

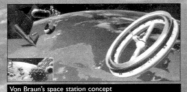
Von Braun's space station concept

Workers inspect Redstone

Von Braun poses with Mars lander model

Nike-Ajax missile

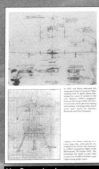
Von Braun sketches

Von Braun's concepts for space flight, including a space station, receive their first national coverage in a series of articles in *Collier's* magazine.

> Here is how we shall go to the moon. The pioneer expedition, fifty scientists and technicians, will take off from the space station's orbit in three clumsy-looking but highly efficient rocket ships ...
>
> Thirty-three minutes from take-off we have it. Now we cut off our motors; momentum and the moon's gravity will do the rest.
>
> The moon itself is visible to us as we coast through space, but it's so far off to one side that it's hard to believe we won't miss it ...
>
> The Earth is visible, too — an enormous ball, most of it bulking pale black against the deeper black of space, but with a wide crescent of daylight where the sun strikes it.
>
> —Dr. Wernher von Braun,
> "Man on the Moon: The Journey"
> *Collier's* magazine, October 18, 1952

Mars study illustration

Some scientists, including some members of the Huntsville team, question the propriety of promoting space exploration through the popular press.

> "We can publish scientific treatises until Hell freezes over. But unless we make people understand what space travel is, and unless the man who pays the bills is behind us, nothing is going to happen."
> — Dr. Wernher von Braun comment to Dr. A.K. Thiel quoted by Mike Wright in *Alabama Heritage* magazine, Summer 1998

> ... Dr. Wernher von Braun, is still only 40 and is the major prophet and hero (or wild propagandist, some scientists suspect) of space travel. As a boy, Dr. von Braun wanted to go to the moon. He still does.
>
> The practical rocket men fear that their gradual march toward space may disappoint the public ... But Dr. von Braun ... who would hurry the cautious missile men along, says that manned space flight "is as sure as the rising of the sun."
>
> — *TIME* magazine, December 8, 1952

Concept of Mars landing

The Provisional Redstone Arsenal Ordnance School is created to teach missile technology to soldiers. This school is the forerunner of what is now the U.S. Army Ordnance, Munitions and Missile Center and School at Redstone Arsenal.

At The University of Alabama's Huntsville center, Dr. Wernher von Braun presents a current affairs seminar, "A Proposed Trip to the Moon."

Model of manned Mars spacecraft

1953

Von Braun shows off the V-2 model

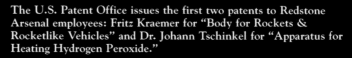
First Redstone launch

Von Braun visits with industrial contractors

A rocket is prepared for transport

In August, the first successful launch from Cape Canaveral, Florida, of a Redstone missile incorporates the first inertial guidance system developed in the U.S.

The U.S. Patent Office issues the first two patents to Redstone Arsenal employees: Fritz Kraemer for "Body for Rockets & Rocketlike Vehicles" and Dr. Johann Tschinkel for "Apparatus for Heating Hydrogen Peroxide."

The first Hermes guided missile is put on display in Huntsville as part of Armed Forces Day.

Philip Mason is named director of The University of Alabama's Huntsville Extension Center.

Hermes test missile

Redstone rocket sits in hangar

1954

Von Braun explains details of Mars craft

Plans made for Disney program

Von Braun demonstrates 'bottle suit'

Walt Disney, Wernher von Braun

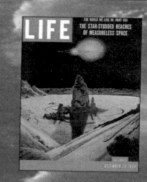

In September, through the Department of the Army, von Braun submits to the U.S. Committee for the International Geophysical Year the report, "A Minimum Satellite Vehicle Based on Components Available from Missile Developments of the Army Ordnance Corps." It proposes using the Redstone missile as the main booster of a four-stage rocket. This concept later becomes the joint Army-Navy Project Orbiter, and still later the Explorer 1 satellite. The Redstone rocket project survives despite President Dwight Eisenhower giving responsibility for launching the first U.S. satellite solely to the Navy.

> Will man ever go to Mars? I am sure he will, but it will be a century or more before he's ready. In that time, scientists and engineers will learn more about the physical and mental rigors of interplanetary flight and about the unknown dangers of life on another planet. Some of that information may become available within the next 25 years or so, through the erection of a space station above the Earth ... and through the subsequent exploration of the moon . ...
>
> — Dr. Wernher von Braun
> "Can We Get to Mars?"
> Collier's magazine, April 30, 1954

First contacts were made with Dr. James van Allen, with the suggestion that he should prepare a scientific experiment for the Explorer 1 satellite.

On November 11, the first 41 members of the team were awarded American citizenship.

Dr. Wernher von Braun becomes a national television spokesman for space exploration, working with Walt Disney to develop three programs which air in 1955 and 1957. Disney's objective, he said, was to combine "the tools of our trade with the knowledge of scientists to give a factual picture of the latest plans for man's newest adventure." It also gives Disney much needed material for the Tomorrowland section of Disneyland.

Rocket team members gain U.S. citizenship

Corporal rocket

Gen. Toftoy, Wernher von Braun

1955

Rocket team members become American citizens

Von Braun consults with technician on a Redstone rocket

Redstone rocket launch

Missile test stand at Redstone Arsenal

On July 29, President Dwight Eisenhower announces that the U.S. will launch a man-made satellite as a contribution to the International Geophysical Year. All three military services submit alternative plans for accomplishing the satellite launches.

U.S. TO LAUNCH EARTH SATELLITE

This country plans to launch history's first man-made, earth-circling satellite into space during 1957 or 1958.

Tentative plans envision an unmanned globular object about the size of a basketball. The satellite will flash around the earth about once every ninety minutes at a speed of 18,000 miles an hour in a fixed path 200 to 300 miles above the ground.
— *The New York Times*, July 30, 1955

The first satellite may have a mouse aboard, but scientists said they could not foresee the time when human beings would be able to go into outer space as passengers ...

Defense Department research scientists pleaded with reporters to repress any tendency to exploit speculations that have been popularized in recent years by fiction writers. They said the possibility of human passengers in a man-made satellite and its use for military purposes were so remotely in the future that speculations about it were practically useless.
— *The New York Times*, July 30, 1955

The first two of three television programs from the collaboration of Walt Disney and Wernher von Braun are broadcast in March and December. An estimated 42 million people see the first program, "Man in Space."

On April 14, a group of 103 former members of the German rocket team and their families receive U.S. citizenship. Among those taking the oath is Dr. Wernher von Braun.

Designed by von Braun rocket team member Hannes Luehrsen, Huntsville's Memorial Parkway opens to traffic. At this time, Memorial Parkway is a bypass around the congested courthouse square in downtown Huntsville.

Huntsville celebrates the sesquicentennial (150th anniversary) of its founding by John Hunt.

The faculty at The University of Alabama's Huntsville center grows to 21 part-time instructors.

1956

Gen. John Medaris

Gens. Medaris, Toftoy

ABMA scientists, from left: Dr. Ernst Stuhlinger, Dr. H. Hoelzer, K.L. Heimburg, Dr. E.D. Geissler, E.W. Neubert, Dr. W. Haeussermann, Dr. Wernher von Braun, W.A. Mrazek, Eberhard Rees, Dr. Kurt Debus and H.H. Maus

On Sept. 20, a Jupiter C vehicle — a Redstone missile first stage with two solid rocket cluster upper stages — achieves a deep penetration of space, reaching an altitude of more than 682 miles and a range of 3,355 miles. With one more solid propellant rocket on top, this rocket could have launched an American satellite in 1956.

The Huntsville team is so close to launching a satellite (an honor reserved for the Navy's Vanguard program) that the Pentagon orders the Army to launch only dummy payloads. At each launch, a Department of Defense official is on hand to make sure the Army missile's fourth stage rockets are not fired — on purpose or "accidentally."

The Ordnance Missile Command becomes the U.S. Army Ballistic Missile Agency (ABMA), with Major Gen. John Medaris as commander and Dr. Wernher von Braun as director of the Development Operations Division. ABMA's mission is to develop, build and field an intermediate range (up to 1,500 miles) ballistic missile, the Jupiter. The Chrysler Corporation was contracted to produce the Jupiter missiles.

The first Redstone missile battalion is activated at Redstone Arsenal and attached to the U.S. Army Ballistic Missile Command.

Redstone rocket launch

Canaveral test stand

Rocket City, U.S.A.
Going Behind the Riddle of Redstone Arsenal

The yellow, one-story building sprawls gracelessly in the Alabama sun. Yet an inconspicuous sign identifies this one-time warehouse as one of the most important buildings in the world. "Guided Missile Development Division," the sign says. This is where, next week, the new Army Ballistic Missile Agency will begin a "crash" program to produce a guided missile with a 1,500-mile range.

Chief of the Guided Missile Development Division at Redstone is a 43-year-old genius named Dr. Wernher von Braun, who directed development of the German V-2 and led the ablest of his staff to the U.S. after the war ... U.S. authorities consider him so valuable that when he returned to Germany to marry his childhood sweetheart, they assigned an Army company to protect him against possible kidnapping by the Reds ... When the Germans arrived, Huntsville was a friendly, busy little city ... At first, Huntsville was wary of the German newcomers.

But, as the newcomers bought land and built houses, proved friendly neighbors and good credit risks, the tensions relaxed. Huntsville citizens began to take pride in the fact that if a power motor wouldn't start, you could call over a world-famous rocket expert from next door to see what was wrong ... This year the new citizens will vote for the first time in an American national election.

"They are respectful of politicians," says Mayor (R.B. "Speck") Searcy, with a certain amount of awe in his voice. "They must have been brought up to respect the "burgermeister."

— *Newsweek* magazine, January 30, 1956

At The University of Alabama's Huntsville center, Dr. Dennis Nead becomes full-time director of graduate programs in physics, engineering, mathematics and management. Upper level and graduate courses in mathematics, physics and engineering are offered in response to the continuing need for technical personnel at Redstone Arsenal. Funding in the early years is provided by the U.S. Army at Redstone Arsenal and later by NASA.

Hermes rocket display

Test stand on Redstone Arsenal

ABMA test stand

The construction crew lifts the observatory dome into place

Drs. Ernst Stuhlinger and von Braun at Monte Sano observatory

1957

Russians launch Sputnik

Life features the von Braun family

President Eisenhower poses with Jupiter nose cone

Von Braun Astronomical Society

Vanguard missile

Jupiter launch

Von Braun astride a Redstone rocket

Rocket test at Redstone Arsenal

The first successful firing of a 1,500-mile-range Jupiter missile from the Atlantic Missile Range is made at Cape Canaveral. The first man-made object recovered from space is a Jupiter C reduced-scale nose cone, proving that Army Ballistic Missile Agency (ABMA) has met the challenge of developing a vehicle that can survive re-entry. President Eisenhower shows off the nose cone during a nationally televised speech from the White House. It later is placed on permanent exhibition in the Smithsonian Institution.

> On the night of Oct. 4, 1957, von Braun was called away from a Redstone dinner honoring Defense Secretary-designate Neil McElroy.
>
> > Voice on the wire: "New York Times calling, doctor."
> > Von Braun: "Yes?"
> > Timesman: "Well, what do you think of it?"
> > Von Braun: "Think of what?"
> > Timesman: "The Russian satellite, the one they just orbited."
>
> Von Braun hurried back to the dinner table, broke the news of Sputnik I, turned earnestly to Neil McElroy. "Sir," he said, "when you get back to Washington you'll find that all hell has broken loose. I wish you would keep one thought in mind through all the noise and confusion: we can fire a satellite into orbit 60 days from the moment you give us the green light." Army Secretary Wilber Brucker, who had accompanied McElroy, raised a hand of objection: "Not 60 days." Von Braun was insistent: "Sixty days." General Medaris settled it: "Ninety days."
> … two weeks after taking office (McElroy) made his decision.
> — *TIME* magazine cover story, "Missileman Von Braun," Feb. 17, 1958

With the president's approval the Department of Defense directs the ABMA to use a Jupiter C rocket to boost into orbit a satellite designed by the Jet Propulsion Laboratory and Dr. James van Allen from the University of Iowa.

Huntsville, Redstone Arsenal and the ABMA rocket team begin a remarkable sprint into history.

> The very next week, (von Braun) reserved Cape Canaveral range time for the night of Jan. 29, 1958, between 10:30 p.m. and 2:30 a.m. Jupiter C had been ready for months.
>
> Says von Braun: "All she needed was a good dusting …"
>
> But the satellite itself, with its delicate instrumentation, might well have held the whole project up for months or years — had not Wernher von Braun, during most of the period that he was barred from engaging in satellite work, been in what he calls "silent coordination" with Caltech's William Pickering and the University of Iowa's James Van Allen in planning Explorer and its instruments.
> — *TIME* magazine, Feb. 17, 1958

Three years toward the completion of an undergraduate degree can be completed without leaving The University of Alabama Huntsville Center, which publishes its first student newspaper, "The Twilight Times."

1958

Charles Lundquist briefs von Braun, Hermann Oberth

Jupiter ballistic missile

Drs. Pickering, van Allen, von Braun celebrate first U.S. satellite launch

Gen. Medaris anticipates Explorer I launch

Redstone rocket undergoes check-out

Jupiter C heads to Cape Canaveral

After many hard years of work, on January 31, 1958, using the Jupiter C — a modified Redstone rocket— von Braun and his team put the first American satellite, Explorer 1, into orbit. Von Braun says of this accomplishment: "It was one of the great moments of my life. I only regret we didn't do it earlier."

VOYAGE OF THE EXPLORER

A bright, waxing moon rode through the racing cumulus clouds above Florida's Cape Canaveral.

Only the week before, the Navy had babied its slender satellite-laden Vanguard. Day by day the tension tightened as the Vanguard countdown crept steadily toward zero ... Each time the launching was scrubbed ...

In the Pentagon at that moment, Army Secretary Wilber Brucker and the Jupiter's top scientist ,Wernher von Braun, joined a score of other military and civilian officials in the Army's telecommunications room ... Elaborately, von Braun lectured the attending brass ...

Then: 10:48, T minus zero — Jupiter C was fired ... After 14 seconds, flame belched from the rocket base ...

Fire Dance. The rocket moved. The Pentagon teletype sang: IT'S LIFTING ... IT'S SOARING BEAUTIFULLY ... Like a flame-footed monster it kicked upward ...

— *TIME* magazine, February 10, 1958

The first American satellite, carrying Geiger counters for cosmic ray measurements prepared by Dr. van Allen, discovers belts of radiation encircling the Earth.

At Redstone Arsenal, the Army team's success in space exploration builds upon itself. The Army's first lunar trajectory probe, Pioneer III, reaches an altitude of 66,654 miles on its way to the moon. An Army Jupiter C launches Explorer 4 to measure artificial radiation belts generated by the explosions of three nuclear bombs launched aboard Redstone rockets high above the Pacific Ocean.

With funds provided by the Advanced Research Projects Agency, von Braun initiated the rocket development program that led to the Saturn-Apollo moon landing project. The first steps in this program were the development of the huge F-1 rocket engines, five of which were to propel the Saturn V rocket, and the development of engine swiveling and engine clustering technologies. These were first studied with Saturn I and Saturn IB rockets before being used on the Saturn V.

The year also sees the first firing of a Redstone missile by combat troops at Cape Canaveral. A full-scale Jupiter intermediate-range ballistic missile nose cone is recovered. Jupiter missiles are delivered to the U.S. Air Force for overseas deployment.

Congress creates the National Aeronautics and Space Administration (NASA).

The city of Huntsville gives 83 acres of farmland to The University of Alabama's Huntsville center. The land is to be used for developing a university campus.

Crane lowers Explorer I into Jupiter C rocket

Jupiter C seconds before launch

Moon Day celebration

Huntsville leaders attend first satellite launch

1959

Jupiter rocket greets Redstone Arsenal visitors

Von Braun talks to Mercury astronauts

'Miss Baker' and 'Able' made their debut for the press and in space

President Dwight Eisenhower presents the Distinguished Federal Civilian Service Award to Dr. Wernher von Braun.

"Practicality, what wonders have you denied man?" von Braun asks at a ceremony honoring Dr. Robert Goddard. "We must open our vision to the unknown. We must expect the unpredictable. We must value knowledge for its own worth, and we must cease to measure the new in terms of its usefulness along."

The Army's success in space exploration continues with the first successful, suborbital space launch and recovery of two monkeys, Able and Miss Baker, on May 28. They are the first space travelers to return safely to the Earth.

A Juno II, the second Army vehicle to launch a NASA probe for lunar exploration, lofts Pioneer IV onto a trajectory past the moon and into orbit around the sun. Explorer 6 returns the first television views of Earth. Explorer 7, also launched by a Juno II rocket, measures the Earth's magnetic field and observed solar flares.

Von Braun and astronaut John Glenn discuss space vehicle model

Redstone's Geniuses Prepare to Go Civilian

One of the richest prizes in all the feuding over the U.S. space and missile program has long been the Development Operations Division of the Army Ballistic Missile Agency at Huntsville, Ala. Stationed since 1950 at the Redstone Arsenal, the division has never been noted for possessing a gaudy array of missile development equipment. For the brilliance of its staff, however, it has gained worldwide recognition ...

Now a change is to come into Redstone's life. The Army's role in the space program has been increasingly controversial, and last October President Eisenhower announced plans to transfer the Development Operations Division to civilian hands — the National Aeronautics & Space Administration.

"Up to now," says von Braun, "the group's bread-and-butter jobs have been military."

His men have sneaked in work on a few non-military projects, such as the Jupiter C and the Explorers, by various bootlegging techniques. "Now," von Braun adds, "we have the chance to go honestly after the main objective. The team has been waiting for this moment for many years."
— *Business Week* magazine, Nov. 28, 1959

Redstone rocket displayed at Smithsonian

Technicians man Pioneer IV launch console

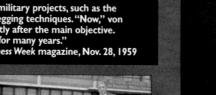

Von Braun receives the Distinguished Federal Civilian Service Award

Redstone rocket launch

1960

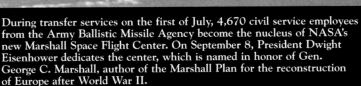
Saturn 1B undergoes static fire testing at Redstone Arsenal

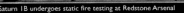
ABMA employees are transferred to NASA

Von Braun explains rocketry to Eisenhower

During transfer services on the first of July, 4,670 civil service employees from the Army Ballistic Missile Agency become the nucleus of NASA's new Marshall Space Flight Center. On September 8, President Dwight Eisenhower dedicates the center, which is named in honor of Gen. George C. Marshall, author of the Marshall Plan for the reconstruction of Europe after World War II.

"I find that the leaders of the new space science feel as if Venus and Mars are more accessible to them than a regimental headquarters was to me as a platoon commander 40 years ago. To move conceptually, in one generation, from the hundreds of yards that once bounded my tactical world to the unending millions of miles that beckon these men onward, is a startling transformation."

– President Dwight Eisenhower

Test of Mercury/Redstone escape system

The first full, eight-engine static firing of a Saturn I booster is conducted in Huntsville.

Pioneer 5, launched atop a Juno II, measures radiation and magnetic fields between Earth and Venus.

The space race between the U.S. and the Soviet Union accelerates.

TWO SOVIET DOGS IN ORBIT RETURNED TO EARTH ALIVE AS SATELLITE IS RETRIEVED

MOSCOW - Living creatures have returned safely to earth from an orbit in space for the first time in history, the Soviet Union announced today.

It said its "second cosmic space ship" landed on target today after circling the earth for twenty four hours with its cargo of two dogs, some rats and mice, flies, plants, seeds and fungi.

— *The New York Times*, August 21, 1960

President Eisenhower visits Huntsville

Huntsville's Lyric Theatre hosts the premier of the movie, "I Aim at the Stars," featuring the life story of Dr. Wernher von Braun, starring Curt Jurgens in the title role.

Huntsville, Madison County and The University of Alabama's Huntsville Center each contribute $250,000 toward the construction of a central building on the new campus. Morton Hall is named in honor of Dr. John Morton, the center's first acting director.

Saturn 1B, Juno II sit in test stand

Launch control at Cape Canaveral

Army exhibits missile systems

Air Force training on Jupiter rocket

1961

Huntsvillians watch static test

Saturn I is hoisted onto test stand

Armed Forces Day - Huntsville

Saturn I sits in test facility

Saturn booster rockets are fired

First launch of Saturn I

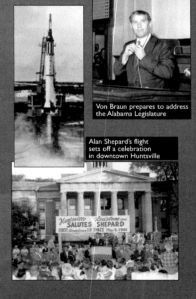

Von Braun prepares to address the Alabama Legislature

Alan Shepard's flight sets off a celebration in downtown Huntsville

The first flight test of a Saturn I booster is successful.

The Soviet Union launches Major Yuri Gagarian into Earth orbit. The flight of the first man in space is a reminder that the U.S. space program is still behind.

> **MOSCOW - The Soviet Union announced today it had won the race to put a man into space ...**
> — *The New York Times*, April 12, 1961

On May 5, Alan B. Shepard, Jr., becomes the first U.S. astronaut in space, launched atop a Mercury Redstone vehicle on a suborbital flight. In the months leading up to the launch, Marshall Space Flight Center scientists, engineers and technicians obsess about making sure everything works as planned. "We could not get it out of our heads that there was a man in that spacecraft," says Dr. Ernst Stuhlinger.

> **CAPE CANAVERAL, Fla. - A slim, cool Navy test pilot was rocketed 115 miles into space today.**
>
> **Thirty-seven-year-old Cmdr. Alan B. Shepard Jr. thus became the first American space explorer.**
>
> **Commander Shepard landed safely 302 miles out at sea fifteen minutes after the launching. He was quickly lifted aboard a Marine Corps helicopter.**
>
> **"Boy, what a ride!" he said as he was flown to the aircraft carrier Lake Champlain four miles away.**
> — *The New York Times*, May 6, 1961

"Our opponents across the ocean, behind the Iron Curtain, thought about a month ago they had slammed the door to the universe in our face, but Shepard has let us out of our dilemma and embarrassment.

"We will go farther and farther, eventually landing on the moon."
> — Wernher von Braun
> public celebration of Freedom 7 flight
> Courthouse Square, Huntsville

On May 25, with a single manned spaceflight under NASA's belt, President John F. Kennedy raised the stakes: "I believe that this nation should commit itself to achieving this goal, before this decade is out, of landing a man on the moon and returning him safely to the Earth."

Alabama deals with desegregation. A leader on so many fronts, von Braun joins others urging a progressive, anti-racist stance, a position taken to heart by many.

> **"We aren't Yankees yet, mind you, but if the governor tried to shut our schools to keep the colored out, he'd be badly mistaken. Huntsville just won't stand for it."**
> — Fifth generation Huntsvillian,
> quoted in *U.S. News & World Report*, 1961

Von Braun addresses the Alabama Legislature, requesting funds to build and equip a research institute on the university campus, saying:

"Opportunity goes where the best people go, and the best people go where good education goes. To make Huntsville more attractive to technical and scientific people across the country — and to further develop the people we have now — the academic and research environment of Huntsville and Alabama must be improved and improved immediately.

"It's the university climate that brings the business. Let's be honest with ourselves. It's not water, or real estate, or labor or cheap taxes that brings industry to a state or city. It's brain power. Nowadays, brainpower dumped in a desert will make it rich."

The legislature approves $3 million in revenue bonds. Huntsville and Madison County buy an additional 200 acres for the campus and the Research Institute is built.

1962

Saturn booster prepares for trip to Cape

President Kennedy, VP Lyndon Johnson visit Marshall

GLENN ORBITS EARTH 3 TIMES SAFELY; PICKED UP IN CAPSULE BY DESTROYER; PRESIDENT WILL GREET HIM IN FLORIDA

CAPE CANAVERAL, Fla. - John H. Glenn Jr. orbited three times around the earth today and landed safely to become the first American to make such a flight.

The 40-year-old Marine Corps lieutenant colonel traveled about 81,000 miles in 4 hours 56 minutes before splashing into the Atlantic at 2:43 P.M.
— *The New York Times*, Feb. 21, 1962

Von Braun watches Saturn launch from Cape

On March 23, von Braun's 50th birthday is commemorated by a book, "From Peenemunde to Outer Space," containing technical and scientific papers by his associates. Among the authors is Dr. Rudolf Hermann, first director of the Research Institute at The University of Alabama's Huntsville center. One of the editors is Dr. Ernst Stuhlinger, who later becomes the Marshall Space Flight Center's associate director for science, a UAH researcher and one of von Braun's biographers.

Following is from the transcript of a briefing for President Kennedy given by Dr. von Braun at Huntsville, Ala., on September 11, 1962:

The President: What is the time schedule for each of these?
Dr. von Braun: This [C-1] vehicle has performed two successful flight tests with the first stage alone. The third one, as I said, is under way. The fourth one will be static-fired today ...

John Glenn steps into Mercury capsule

The advanced Saturn will be ready to fly the lunar mission in approximately 1967, if everything works very well. But we like to keep a little time-padding for ourselves, too, so I think this statement should be taken with a grain of salt.

But we will definitely do it in this decade.
— *U.S. News & World Report* magazine, October 1, 1962

At Cape Canaveral, NASA's Launch Operations Center is created from a former element of Marshall Space Flight Center. The director is Dr. Kurt H. Debus, a long-time von Braun associate and former Huntsville resident.

Von Braun helped plan Cape Canaveral facilities

Across from the The University of Alabama's Huntsville center, Cummings Research Park is being developed. Von Braun insists the group developing the park form a non-profit foundation, which later becomes The University of Alabama Huntsville Foundation. National companies come to Huntsville to support the space program.

Federal grants totaling $114,687 are awarded to the Research Institute to support graduate studies.

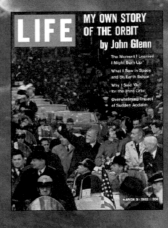
LIFE
MY OWN STORY OF THE ORBIT
by John Glenn

Saturn shown in scale to Statue of Liberty

Saturn components were regular cargo on 'Super Guppy' aircraft

Fifty-year-old Wernher von Braun displays missile development legacy

1963

Von Braun makes supersonic flight at Edwards AFB

President John Kennedy visits Marshall Space Flight Center

The Mercury program ends with the successful flight of Gordon Cooper. Preparations for the Gemini program continue.

On May 19, President John F. Kennedy visits Redstone Arsenal and the Marshall Space Flight Center for a tour and briefings on Project Apollo. He repeats his pledge that the U.S. will explore space:

"I know there are lots of people now who say, 'Why go any further in space?' When Columbus was halfway through his voyage, the same people said, 'Why go on any further? What can he possibly find? What good will it be?'

"I believe the United States of America is committed in this decade to be first in space, and the only way we are going to be first in space is to work as hard as we can here and all across the country ..."

Building 4200, Marshall Space Flight Center's headquarters, is completed.

Von Braun serves as spokesman for space travel

In January, von Braun becomes a monthly contributor to *Popular Science* magazine. On the splashy front cover, beneath a headline that asks, "10,000 miles Between Grease Jobs? How Safe Is Extended Lubrication?" the magazine trumpets: "Starting This Month: Dr. Wernher von Braun Answers Your Questions about Space"

"Space science isn't like geography, or astronomy, or physics, or chemistry, or medicine. It is a little bit of all of them and more. That is what makes it so fascinating.

"But it is this kaleidoscopic aspect of space science that makes it almost impossible to 'organize' a monthly column such as this. Mr. Crossley and I have therefore agreed not even to try to arrange the questions and answers in any systemic way. If the result is a bit disjointed, it should at least be colorful."

— Wernher von Braun
"Why I am writing for *Popular Science*"
Popular Science magazine, January 1963

Von Braun prior to flight in a T-38 aircraft

H. Clyde Reeves, vice president for Huntsville affairs, and Dr. Charley Scott, the first director of instruction, oversee rapid growth of the University of Alabama's Huntsville campus.

Saturn hardware awaits next step of assembly

Kurt Debus, von Braun anticipate launch at Cape Canaveral

F-1 ENGINE SYSTEMS TEST STAND - HUNTSVILLE Aug. 23, 1963
Construction of West Test Area at Marshall Space Flight Center

1964

Von Braun studies data at Cape Canaveral

Dynamic test facility at Marshall Space Flight Center

Engineers inspect Gemini capsule

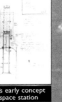
Von Braun's early concept for Skylab space station

Saturn V's first stage is lowered into test facility

The Dynamic Test Facility is built for testing Saturn rockets. It was the tallest structure in Alabama for years.

Q. Dr. von Braun, is it still important to go to the moon?

A. It is just as important as it ever was. The real purpose of going to the moon is, of course, not to go there to put up a sign, "Kilroy was here," or maybe bring back a handful of lunar sand. The purpose is to develop a true, national space-flying capability.

Q. Why is that so important?

A. Because it will vastly extend our knowledge about the universe, and because it will enable us to develop new technologies from which we are bound to reap many benefits in our everyday lives.

Q. But the moon is ... hardly important to everyday life.

A. Look at the moon as more of a rallying point than an objective in itself. The moon plays the same role in our manned space-flight program that the city of Paris played in Lindbergh's memorable flight. Paris served as a goal, but surely Lindbergh had a more important objective in mind than to go to Paris.

Our objective is to develop a broad, manned space-flying capability. You simply cannot develop such a capability ... unless you have a clear goal — a focusing point like Lindbergh's Paris — something that is universally understood ...

Now, when the president of the United States says, "Let's land a man on the moon in this decade and bring him back alive," then you have such a clear goal. Everybody knows what the moon is, everybody knows what this decade is, and everybody can tell a live astronaut who returned from the moon from one who didn't.
— *U.S. News & World Report*, June 1, 1964

With an enrollment of 208 students, the first full-time freshman class begins a regular full-day teaching program at The University of Alabama's Huntsville center. The Huntsville center awards its first master's degree to Julian Palmore. Total enrollment at the center tops 2,000 part-time students. With an enrollment of about 900 students, the center's graduate program is reported to be the largest in the South.

Marshall officials watch launch from Cape

Missile and Space Museum at Redstone Arsenal

Saturn I is launched from Cape Canaveral

Saturn is loaded onto barge on the Tennessee River

1965

Strategic Planning Group: Dr. Ernst Stuhlinger, William Mrazek, Dr. Wernher von Braun, Dr. William Lucas, Ray Kline and Jim Daniels

Space walks added drama to manned space flight

F1 engine – Workhorse of the Saturn V launch vehicle

A Saturn I rocket is launched in February carrying Pegasus I, a five-ton micrometeroid sensing satellite. Pegasus 2 is launched in May and Pegasus 3 in July. All three flights were successful, providing information on the size and frequency distribution of micrometeroids in space.

The Apollo Applications Program is initiated to extend the use of Apollo and Saturn hardware.

Plans for Skylab, the first large U.S. space station, are laid out by the Army Ballistic Missile Agency and NASA.

The Strange World of Zero Gravity
by Dr. Wernher von Braun

When a human astronaut becomes a human satellite, floating freely in the vastness of space, he offers a dramatic example of the weird things that can happen in the realm of zero gravity — a strange world just beginning to be explored ...

Living and housekeeping under zero gravity ... pose curious problems: Before going to sleep, astronauts will need to strap themselves down, or they would float about the cabin, propelled by the thrust of their own breath.

Even the most luxurious space cuisine will avoid needless food wastes: All meat will be boneless and free of undesirable fat; potatoes will be peeled in advance; cherries will be stoneless. If a Martini is ever served aboard a spacecraft, it may come without an olive.

— *Popular Science* magazine, June 1965

First stage of Saturn 1B

On August 5, von Braun interrupts a meeting so he can watch from the roof of Building 4200 as the Saturn V booster's five engines undergo their first full-duration firing. Windows rattle across North Alabama and some plaster cracks. NASA employees a mile or more from the test stand are buffeted by the blast wave.

During earlier, shorter tests in April and May, some frightened area residents wonder if a nuclear bomb has been dropped. It is the middle of the Cold War. Redstone Arsenal is routinely on the Soviet Union's top ten "hit list."

Saturn engines are test fired

"... a continuous plume of flame blasts from the base of a mammoth concrete structure," Marshall Space Flight Center Historian Mike Wright says in an article for *Alabama Heritage* magazine. "Thunder rolls. Smoke billows. For two and a half minutes hell unfolds. Alabama has become, as writer Bob Lionel later wrote, 'the land of the Earth-shakers.' "

In November, the state legislature approves $1.9 million in bonds to finance construction of the Alabama Space & Rocket Center.

"If our education system can instill in our youth an early desire for perfection, it has taken a major step in training to meet the space age challenge."

— Wernher von Braun, April 1965 comments to Alpha Phi Alpha fraternity

Saturn 1B looms tall on test stand

The first named scholarship exclusively for the University of Alabama's Huntsville campus is designated by the will of Octavia May Palmer to honor the memory of her father, Samuel Palmer. The gift includes land in Madison County which has been in the Palmer family for more than 100 years.

Launch personnel observe Saturn I launch

1966

Saturn IB rocket sits in launch mode

Saturn IB is bathed in spotlights at the Cape

Saturn V rests in Vertical Assembly Building

Rocket motors undergo test at Redstone

Saturn ascends into space following launch

The first flight test of the Saturn I-B with a Saturn IV-B upper stage (liquid oxygen and hydrogen propellants) and an Apollo capsule recovery.

To promote space exploration, Dr. Wernher von Braun continues to work with the press. This results in a variety of coverage, including this question-and-answer session with *U.S. News & World Report* magazine:

Q. Beyond (the objective of reaching the moon), what is our purpose?

A. I think that even when President Kennedy made that announcement, he made it perfectly clear that, while this is an objective that everybody understood and one that we could use to construct a program for fulfillment, what he really had in mind was to develop a broad, national space-flying capability. He put it very well when he said, "We have to learn to sail on the new ocean of space."

The problem is, when you have an objective as vague as learning to sail on a new ocean and you try to present this, say, to the Congress, then it isn't surprising that everybody interprets this on his own terms.

What does that mean — to develop a capability to sail on an ocean? There's no objective; there's no goal in space, and no goal in time attached to it.

Q. Getting back to the moon program: Do you think it is so important for us to go there, at such great cost?

A. My firm conviction is that our space-flight program ... (has) helped tremendously to invigorate our sciences, our educational system and our industry. And I think the question is not so much: "Can we afford the space program?" It is: "Can industrial countries which do not have a space program continue not to have one?" — because they are falling hopelessly behind in technologies that spin out of the space program.

Q. Weren't there some disappointments in the Gemini flights?

A. If, in a pioneering program of this magnitude, you run into a surprise, you should not call it a disappointment. You see, the purpose of Gemini was to learn the kinds of things that you can't learn on the ground. And the process of learning inevitably produces surprise discoveries. And we did make some surprise discoveries, for which we are more than grateful, because now we know more than we did when we went into it.

— *U.S. News & World Report*, December 12, 1966

Madison Hall is built with $900,000 raised by the Huntsville community and a $420,900 federal grant. It is called the Graduate Studies Building, but is later renamed to honor the citizens of Madison County.

Concepts for lunar rover

Officials raise money to build Madison Hall on UAH campus

1967

Apollo astronauts Gus Grissom, Ed White and Roger Chaffee

Saturn V in Vertical Assembly Building at Cape Canaveral

Von Braun conducts research in the forbidding Antarctic region

Astronauts Gus Grissom, Ed White and Roger Chaffee die when fire guts their Apollo spacecraft at Kennedy Space Center in Florida in January. Marshall Space Flight Center Deputy Director Eberhard Rees is appointed to chair the group that initiates corrective actions.

The first launch of a Saturn V rocket is on November 9, when a Saturn V rocket carries the unmanned Apollo 4 spacecraft into Earth orbit. "No single event since the formation of the Marshall Center in 1960 equals today's launch in significance," says von Braun. "For MSFC employees — more than 7,000 strong — this is their finest hour."

NASA faces budget cuts. Years later, von Braun says that while it knew budget cuts would slow work in the Apollo program "... Congress was even more aware that there were simply not enough funds to satisfy all the requirements of Vietnam and the many urgent demands for domestic programs."

The University of Alabama's Huntsville center becomes a branch of The University of Alabama System and is officially named The University of Alabama in Huntsville (UAH). UAH's Research Institute receives funding from NASA's Marshall Space Flight Center to prepare the official history of the Saturn Program, "Stages to Saturn."

Huntsville's population in 1967 tops 125,000, a more than sevenfold increase since 1950.

> "I used to be able to say the name of everyone I met on the street — and if they had a dog with them, I knew the dog's name, too. Now it's different."
>
> — Huntsville Major R.B. "Speck" Searcy,
> quoted in *Fortune* magazine

Von Braun monitored details of every element of the space program

Saturn V seconds after liftoff

1968

Earthrise – Apollo 8 astronauts show a different perspective of Earth

Satellite communication was critical for space program

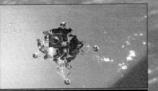
Dr. Kurt Debus (center, foreground), and staff at Launch Complex 37

Astronauts run initial test of lunar module

Command module undergoes 'dunk' tests

Von Braun experiences weightlessness in a KC-135 and in a neutral buoyancy tank

After successful unmanned flights of Apollo 5 and 6, Apollo 7 is launched by a Saturn I-B rocket, with astronauts Wally Schirra, Walt Cunningham and Donn Eisele. Their Apollo capsule lands and is recovered safely after 163 Earth orbits.

A three-stage Saturn V booster launches the Apollo 8 mission, which circles the moon. On Christmas Eve, the three astronauts send a television broadcast from lunar orbit back to the Earth.

As the telecast neared its end, Colonel (Frank) Borman said, "Apollo 8 has a message for you." With that, Major (William) Anders began reading the opening verses from the Book of Genesis about the creation of the earth.

"In the beginning," Major Anders read, "God created the heaven and the earth. And the earth was without form and voice; and darkness was upon the face of the deep ..."

Captain (James) Lovell then took up with the verse beginning, "And God called the light day, and the darkness He called night."

Colonel Borman closed the reading with the verse that read: "And God called the dry land Earth; and the gathering together of the water He called the seas; and God saw that it was good."

After that Colonel Borman signed off, saying: "Good-by, good night, Merry Christmas. God bless all of you, all of you on the good Earth."

— *The New York Times*, December 25, 1968

If all goes well, our goal of landing men on the moon and returning them safely by the end of this decade as planned in 1961 will be fulfilled — that is, if we are not held up by having to pass through Russian customs ...

Speculation on the potential benefits from space made only a few years ago were vague and indefinite. Today they are more visible, credible and attractive to potential users. And yet, while we are largely within the exploration phase of space exploration and still looking forward to the exploitation phase, it is impossible to make predictions with certainty.

History teaches us that the greatest payoffs — the ones which will be the most revolutionary, the most significant and have the widest application — may well come from some unforeseen aspect of the program and from discoveries yet to come.

— Dr. Wernher von Braun in a speech to the National Space Club, quoted in *U.S. News & World Report* magazine, March 18, 1968

Q. Dr. von Braun, is the U.S. going to beat Russia to the moon?

A. *I am beginning to doubt that we will. It is very important that we are there first, but in view of the spectacular performance of the Soviet spacecraft Zond 5, in late September, I am beginning to wonder. It will undoubtedly be a photo finish.*

Q. Does the U.S. have no program beyond a moon landing?

A. *It may surprise you to hear this, but for the last two years my main effort at the Marshall center has been following orders to scrub the industrial structure that we had built up a great expense to the taxpayer, to tear it down again. The sole purpose seems to be to make sure that after 1972 nothing of our capability is left. That's my main job at the moment. And we haven't even put a man on the moon yet.*

— *U.S. News & World Report* magazine, October 14, 1968

1969

"That's one small step for a man. One giant leap for mankind."

MEN WALK ON MOON

HOUSTON - Men have landed and walked on the moon.

Two Americans, astronauts of Apollo 11, steered their fragile four-legged lunar module safely and smoothly to the historic landing yesterday at 4:17:40 P.M.

Neal A. Armstrong, the 38-year-old civilian commander, radioed to earth and the mission control room here:

"Houston, Tranquility Base here. The Eagle has landed."

— *The New York Times*, July 21, 1969

"To be borne by a rocket to the moon is one thing, but to be borne up these steps by admirers is almost as impressive."

— Dr. Wernher von Braun
Apollo 11 celebration
Courthouse Square, Huntsville

The Marshall Space Flight Center begins developing the Lunar Roving Vehicle, which is used by astronauts on the last two Apollo missions, in 1971 and 1972, to explore the moon's surface.

The journey begins...

The Harvest of Operation Paperclip

Recently, Walter Dornberger, who headed German rocket development in World War II, visited the launching complex at Cape Kennedy. "There was an old friend standing as a supervisor at the rocket take-off site," recalls Dornberger, "and I asked him to show me what was really new about the missile. 'Doc,' he replied, 'it is larger, heavier, more reliable, but in the main part it is the same old cucumber.'"

— *Newsweek* magazine, July 7, 1969

If teamwork and a sense of shared responsibility were crucial factors in the U.S. effort to land men on the moon, so were the contributions made by a number of individuals ...

Dr. Wernher von Braun, 57, director of the Marshall Space Flight Center in Huntsville, Ala. ... helped develop the ablative heat shield, which dissipates the searing heat of re-entry by flaking off in harmless fiery pieces. His Huntsville group can also claim credit for what has become known in the space agency as "cluster's last stand" — the grouping of several smaller rockets in a cluster to provide as much thrust as would a single, far larger rocket engine.

... Von Braun, perhaps more than any other man, has been the driving force behind the moon program.

— *TIME* magazine, July 18, 1969

Officials relax after a successful launch

Von Braun, George Mueller confer as launch approaches

UAH's Science and Engineering Building is built. It will later be named in memory of Dr. Harold Wilson, dean of the College of Science.

Construction begins on the first of four planned phases of a permanent UAH library, which is later named in honor of M. Louis Salmon, one of the founders of The University of Alabama Huntsville Foundation and its chairman from 1986 to 1993.

UAH's Student Union Building is built. It includes a bookstore, cafeteria, lecture rooms, auditorium and student organization offices.

Von Braun, Huntsvillians celebrate historic flight

Sen. John Sparkman addresses farewell crowd

Von Braun family: Wernher, Maria, Peter, Margrit and Iris

Von Braun bids a tearful farewell

U.S. Sen. John Sparkman, Alabama Gov. Albert Brewer honor Wernher von Braun

Von Braun photographed in his office at Marshall Space Flight Center

Von Braun passes leadership of Marshall Space Flight Center to Eberhard Rees

History, forever changed — 1950 to 1970

Critics of the National Aeronautics and Space Administration have long maintained that the agency's greatest fault has been its failure to plan for the long-range future after landing men on the moon. ... there can be little doubt that NASA has lost the once-captive imagination of the public. Last week, in an obvious attempt to recoup, NASA announced that it has a new chief planner: Dr. Wernher von Braun, the best known rocket expert in the U.S.

"If we tried to do all of the proposals that are floating around in NASA today ... it would cost the agency anywhere from ten to 100 times its present budget." As the Deputy Associate Administrator for Planning (a newly created post) and the No. 4 man in NASA, von Braun's task is formidable.

— *Newsweek* magazine, February 9, 1970

Dr. Wernher von Braun leaves Huntsville for NASA headquarters in Washington D.C. As deputy associate administrator, he works on space plans for NASA.

"My friends, there was dancing in the streets of Huntsville when our first satellite orbited the Earth. There was dancing again when the first Americans landed on the moon.

"I'd like to ask you, don't hang up your dancing slippers."

– Wernher von Braun
Farewell speech
Courthouse Square, Huntsville

The name of Wernher von Braun is almost synonymous with space exploration ... As director of the George C. Marshall Space Flight Center at Huntsville, Ala., he became the mastermind for the giant Saturn V ...

Today, von Braun has a new job. His task is to define U.S. space goals ...

Q. What types of projects do you think we will be doing by the late 1970s to 1980s?

A. I consider the shuttle the most exciting program that NASA is involved in at the moment. ... the shuttle will reduce transportation costs to orbit. At the moment, it still costs anywhere between $500 to $1,000 to put a pound of payload into a low earth orbit. I think there's reason to believe this cost can be reduced to like $50 to $100 per pound in orbit.

Q. Do you expect ... big, multi-manned space stations in orbit toward the end of the 1970s?

A. Yes, I think so.

Q. What activities will we conduct on the moon in the next decade?

A. By the end of the next decade we are likely to have several small lunar base camps ...

— *Business Week* magazine, July 4, 1970

Dr. Eberhard Rees, a member of the von Braun team, is named as the new Marshall Space Flight Center director.

UAH enrollment reaches 2,173. UAH becomes an autonomous institution and Dr. Ben Graves is appointed the university's first president. UAH has its first commencement ceremony. A graduating class of 130 hears CBS newsman Eric Severeid give the commencement address.

Artist rendering by Tom Fricker of Alan B. Shepard, Jr., commissioned in May 1991 on the 30th anniversary of the flight of Freedom 7 honoring Shepard as the first American in space, Naval Aviator, Commander of Freedom 7 and Apollo 14, Admiral, U.S. Navy and President, Mercury 7 Foundation.

Apollo 17 moon buggy, stripped down for high speed test: driver and last man on the moon (inset), Gene Cernan

"Dr. Rock", Harrison Schmitt stands next to moon boulder at Apollo 17 landing site in the Taurus mountains

Von Braun gives his Huntsville farewell speech on the occasion of his transfer in 1970 to NASA Headquarters, Washington, DC. His wife, Maria, is seated at right.

Von Braun sells space to a wide range of audiences. During a speaking engagement in Oklahoma, he is named Chief-Fire-Arrows-To-The-Moon (right)

Von Braun with a model of the nuclear powered stage of Saturn V that he proposed in 1969 for a manned Mars mission (below)

Shepard serves as honor guard for
famous Geisha in Japan (left)

Lift-off of the Mercury Redstone.
Deke Slayton: "You are on your way Jose!"

EXPERIENCED ASTRONAUT

Roasting Shepard at his 30th anniversary of Freedom
7 Mercury flight. Would you fly to the moon with this
man? (right)

Schirra aboard his Sigma 7 Mercury spacecraft. "Ride that thing ,Wally," shades of Dr. Strange Love (above)

Parade featuring Corvettes and Astronauts at Cocoa Beach with Shepard leading thirty of his friends (left)

The Shepard women from left, Julie Shepard Jenkins, Laura Shepard Churchley, author, Alice Shepard Wackerman and Louise Shepard, at the 30th anniversary of Shepard's Freedom 7 flight in 1991, Washington, DC

U.S. Space Camp logo

The remaining M7-Carpenter, Cooper, Schirra and Glenn with life size sculpture honoring Shepard at Astronaut Hall of Fame

President George H.W. Bush visits twice, once as vice president and the second time as the 41st President of the United States

Thunder in Alabama. The Saturn V moon rocket booster generating 7.5 million pounds of thrust-equal to 160 million horsepower-is shown undergoing a ground test firing in Huntsville

Gotcha by Schirra. Shepard views astronaut's memorabilia thinking all of the Astronaut Hall of Fame collection belongs to Schirra

Shepard speaking at Huntsville Space Camp graduation in 1992 . When Buckbee asked the question, " Who wants to go to Mars?", everyone raised their hand, including Shepard

Official opening of U.S. Astronaut Hall of Fame, Florida, 1990

Arnold Schwarzenegger, currently the Governor of California, is shown here representing the President's Physical Fitness program receiving an honorary astronaut hat and Space Camp medallion from Buckbee

Buckbee left, with Mr. & Mrs. Henri Landwirth and Senator & Mrs. John Glenn at Mercury-Atlas launch site with M7 logo commemorating Mercury flights from Pad 14

M7 and Bill "José Jimenez" Dana, third from left, the eighth Mercury astronaut and he has a lapel pin to prove it.

Earth is the cradle of man but one cannot live in the cradle forever. **Konstantin Tsiolkovsky**

I remember the night of October 4, 1957. I was working on the campus newspaper, *The Daily Athenaeum*, in the West Virginia University School of Journalism. The United Press wire service machine rang more bells then I had ever heard. The U.S. Department of Defense confirmed the Russians had just placed a satellite in outer space, something called a Sputnik. I remember the word vividly because I looked it up in Webster and it wasn't there. Little did I know that night how soon I would be in the midst of America's race to the moon. Four years later, I was standing on the beaches of Cape Canaveral watching Alan Shepard blast off atop a Redstone rocket to become the first American in space. Three weeks later, President Kennedy committed the nation to landing a man on the moon and returning him safely, and doing it before the Russians. It was the beginning of the Space Race and my career as a space marketeer.

The first time I encountered a real, live Russian was in 1970 when Buzz Aldrin brought two cosmonauts, Andrian Nikolayev and Vitaly Sevastyanov, to the Space & Rocket Center. This wasn't many months after Buzz had walked on the moon and the Russians were a bit envious of our accomplishment. Cosmonaut Nikolayev had flown in Vostok 3 in 1962, part of the first Russian crew to fly in formation, bringing the spacecraft within three miles of the Vostok 4. Cosmonaut Sevastyanov joined Nikolayev aboard Soyuz 9 in 1970 for one of the first long-duration missions of 17 days. Upon their return, they were so weak they had to be carried from the Soyuz spacecraft. This was the first indication that cosmonauts and astronauts would not be able to endure long space flights without exercising.

Cosmonauts Vitaly Sevastyanov, Andrian Nikolayev, Apollo 11 astronaut Buzz Aldrin and Buckbee touring museum

The delegation arrived at the Space & Rocket Center with translators and heavy security people, including KGB, FBI, CIA and NASA security. I felt it was an exercise in establishing who could produce the greater number of spooks at one location. Everyone was very formal, almost military-like. Appearances could have led one to believe it was a delegation prepared to conduct peace talks, rather than visit a space museum.

We had on display a movie prop of Vostok with a large red star hanging high from the ceiling of the museum. As we were touring the building, one of the cosmonauts pointed to the Vostok and asked, *"Where did you get that?"* In my quick response I told the translator we recovered it from the mountains in Mongolia. There was an immediate reaction as the translator told the cosmonauts and the KGB what I said. I could tell the Russians were not happy and I quickly added, *"It's a joke, it's a joke; it's a movie prop."* In the interest of avoiding an international incident, I decided not to make any more jokes.

As we walked through the museum and began a tour of the giant Saturn V moon rocket—the only one on public exhibit in the world at that time—the Russian delegation, in particular the cosmonauts, became extremely interested. They asked numerous questions regarding the type of engines, the thrust, and the kind of fuel used. Buzz and I answered in rapid fire as we received the questions. As we were walking along the Saturn V, which was about the length of a football field from booster to the spacecraft, I asked the question, *"What kind of fuel do you use in the upper stages of your rockets?"* There was total silence and the technical guy, the one I had supposed to be a space agency expert, answered, *"I don't know."* The cosmonauts also shook their heads. The exchange of information during that two hour session was entirely one-sided. We provided answers to everything that was asked and I don't recall receiving any information from the Russians to answer our questions. Finally, pictures were taken, hands extended and grasped in firm, cordial grips, and Buzz and I said good-bye to our cosmonaut friends.

My next encounter with Russians was on their turf. I visited six times in the late 1980s and early '90s. I encouraged Shepard to accompany me, but his usual response was, *"The food is bad, the women are ugly and there is nothing to do."* He still looked upon them as the Russkies, the evil empire, and had no desire to develop any relationships. On my first trip, I joined a U.S. delegation consisting of NASA public affairs people, space historians and writers who were attending a worldwide conference in Moscow on the impact of U.S. and U.S.S.R manned spaceflight programs. We toured Star City, the Gagarin Air Force Museum and other aircraft and space museums and exhibitions. One memorable visit was to the museum that had the remains of the U. S. pilot Francis Powers' U2 plane. The Russians were quick to point out this was the remains of the spy plane they were successful in shooting down during the Cold War. It seemed a little odd; here we were, a group of Americans having our picture taken with the remains of an American-built U2 plane shot down by our Russian host.

USSR Star City entrance sign and Buckbee with Lt. General Vladimir Shatalov, director of Cosmonaut training

As the sessions began, our representative was the first to take the podium with a detailed briefing of the Mercury, Gemini and Apollo manned spaceflight programs. The Russians, led by academician B. K. Rauschenbak, gave a similar briefing, stating instead of going to the moon with manned crews, they chose to do more complicated lunar missions with machines and robots. He proceeded to give a detailed explanation of the stellar operation of their unmanned robot. He closed his presentation by saying, *"The U.S.S.R. had accomplished truly scientific exploration of the moon and was continuing to receive scientific data still being studied by the scientific community."* He ended by saying, *"The U.S.S.R. lunar program was truly a scientific accomplishment and the U.S. Apollo program was nothing more than a stunt."*

Up to this point I'd been very considerate of our Russian hosts, but that comment brought me out of my chair. I asked permission to speak and interjected that Rauschenbak's statement disturbed me. I reminded the audience that the U.S. had landed not one but twelve men on the moon at six different landing sites, returned several hundred pounds of lunar samples which we continue to study, and shared samples with scientists throughout the world to conduct their own investigation. We had not only landed on the moon, but astronauts had driven a vehicle on the moon. One astronaut had demonstrated hitting a golf ball, one of the great sports appreciated and followed by Americans. It was eloquently pointed out by another member of our delegation that Russia cancelled their manned moon landing program after two deadly rocket explosions on the launch pad. The Russian gentleman who made the previous statements looked straight at me as I was speaking to him and obviously wasn't happy with my comments regarding the Apollo program. When I sat down there was immediate applause from many young Russians and, of course, the Americans present. The chairman of the panel thanked me and announced he was quite certain the academician didn't intend to discredit or belittle the U.S. Apollo program, but was merely describing the difference between the Soviet and American approaches to exploration.

Buckbee and delegation member view remains of Frances Power's U2 spy plane shot down by the USSR

Cosmonaut Gherman Titov, second man to orbit the earth, is Buckbee's host

I returned to Russia several times during the late '80s and early '90s. The purpose was to acquire Russian space artifacts that would become part of a traveling exhibit of Russian and American space hardware. I also was interested in developing a Space Camp in Russia. These visits brought me in touch with many of the cosmonauts, as well as allowed me access to the factories that built the Russian space hardware. Another high point from my travels was the opportunity to visit and live at Star City where the cosmonauts trained.

I met Gherman Titov, call sign "Eagle," the second man to orbit the Earth after Yuri Gagarin. Now retired, he led a company that would produce replicas of Vostok hardware for the exhibit, but only after a bit of lively negotiation between the two of us. I met cosmonauts Nikolayev, Popovich and Leonov who flew on the 1975 U.S.-Russian Apollo Soyuz mission, and also met the first woman to fly in space, Valentina Tereshkova who was married to Nikolayev.

Soyuz fliers Valentin Lebedev, Alexander Serebrov, Vladimir Lyakhov, Alexander Alexandrov and Vladimir Solovyov came to the U.S. and were guest lecturers at Space Camp. Solovyov's son accompanied him and attended one of the first International Space Camp sessions held in Huntsville. Consequently, I developed a warm relationship with Solovyov. After he retired from active spaceflight, he became an influential official in the Soviet space agency. During one of our visits, we discussed the opportunity for the Coca-Cola Company to fly their specially designed Coke "space can" on a Russian spacecraft. The Coca-Cola Company had flown it on the Space Shuttle and wanted to get some more flight time, having cosmonauts use it in zero gravity. After several months of negotiation, arrangements were made and a six-pack of Coke space cans was flown and demonstrated by the cosmonauts.

Comparing astronauts and cosmonauts was an interesting study. Titov and many of the early cosmonauts were true heroes, having cars, apartments and front row seats at any highly acclaimed event, compliments of the Russian government. Wherever they went, they expected to be seated first, have the best food and pay nothing. Firsthand, I experienced driving on the wrong side of the road for several blocks in Moscow as my cosmonaut host passed numerous cars, trucks and taxis, including two policemen, traveling to a performance of the Russian Ballet. Wherever they ventured, there were people to serve them. All of the early Russian flyers were made Heroes of the Soviet Union and awarded the Order of Lenin. Some received these high awards twice. Even after they retired, they wore their awards on military uniforms.

Star City, home of the cosmonauts, is a very modern facility to Russian standards, located about an hour out of Moscow. Interestingly enough, it resembles NASA's Johnson Space Center in Houston. Unlike Johnson Space Center, however, the Russian space fliers are eulogized within the complex. Yuri Gagarin, first human to orbit the Earth has the highest honors within Star City. The film of his flight is shown to all visitors as if he invented manned spaceflight. The campus-like training center offers cosmonauts pleasant housing and well stocked shops that contrast sharply with domestic products available to average Russians living in Moscow. I had brought some gifts for the cosmonauts, video tapes of *The Right Stuff* and *Top Gun* movies, and had been worried they might not have the proper equipment to play the American format. I was surprised to see they had VCR's, mostly Sony models, designed to play any video format available in the world, so my offerings were a great hit.

During one visit, my escort was cosmonaut Vladimir Lyakhov, a veteran of three long- duration flights and commander of the Soyuz mission that had a problematic return to Earth. He described the mission in detail, explaining that the automatic system failed, requiring the spacecraft to be in orbit for 24 hours with limited activity including no food or water. He brought the spacecraft under manual control to a successful landing.

Cosmonaut crews in Star City were training to fly the Mir Space Station. I crawled all over their 1-g trainer. Surprisingly, the area resembled the Space Camp training center. I was briefed on the station and the new re-supply module that was being tested for future flights. Crew training was underway, involving transferring equipment from the re-supply module to the living quarters. Lyakhov explained much of the cosmonaut personal gear, such as food, drinking water dispenser, exercising devices, and their new "walk in space" space suit. I had an opportunity to suit-up in Russian gear which resembled our Space Shuttle suit, with the exception that the cosmonaut's version featured entry through the backpack. Another innovative piece of equipment was the launch and re-entry couch, which encompassed most of the cosmonaut's body and reclined similar to a lounging chair. It had shock absorbers to ease the impact of touchdown. Cosmonauts fly fully suited at launch and re-entry. The next stop was the hydro-pod underwater simulator or neutral buoyancy facility, which was very much like Marshall Space Flight Center's original neutral buoyancy tank, and had almost the same dimensions. Cosmonauts and astronauts train underwater on Earth to simulate walks in space. On this day, cosmonauts were practicing with portions of the Mir-Soyuz space station.

Next door is Star City's centrifuge facility which the cosmonauts continue to use throughout their training. My guide indicated that their training was more rigorous than the training American astronauts undergo. The next stop was the Cosmonaut Museum, which had a small souvenir shop at its entrance, offering Soviet spacecraft merchandise and a few medallions of famous spaceflights. The escort officer gave me a fascinating guided tour of the museum which began with the early flights and continued through all significant manned Soviet programs. The rooms were lined with display cases containing personal memorabilia of the crews, some dedicated to particular programs such as Vostok, Voskhod, Soyuz and MIR. Hardware was displayed openly and could be closely examined.

There were a number of items presented by American astronauts Frank Borman and Tom Stafford who had visited and presented gifts to various cosmonauts. In one room of the museum, Yuri Gagarin's personal office had been reconstructed. It contained a huge wall map of the world, a desk, filing cabinets and work table. Before a crewed mission, the traveling cosmonauts visit this room and sit at Gagarin's work table and enter their names and missions in a special log. This area is set aside for meditation and remembering the sacrifices their colleague made to the Soviet space program. Star City's private dining room offers some of the best food in all of Russia.

Lieutenant General Vladimir Shatalov, director of cosmonaut training at the Yuri Gagarin Cosmonaut Training Center, hosted one of my visits. Many in the military held important posts in the space program. He was the highest ranking military person I met during my visits to Russia. The three-time space flier and holder of two Hero of the Soviet Union and two Order of Lenin honors briefed me and discussed training plans for future long duration missions aboard space stations. A highly respected engineer and technical wizard, he was quite proud of the fact that he had been involved in the joint U.S.-Russian Apollo Soyuz mission and had visited Johnson Space Center. Deke Slayton had suggested I visit him because they knew each other and had spent time together during Apollo-Soyuz mission training. Unfortunately, after that joint mission in 1975, all communications were cut off by the Russians with any official NASA people and didn't resume for almost 15 years (for reasons only the politicians know).

While at the training center, I met cosmonaut Alexander Alexandrov, a cosmonaut/pilot from Bulgaria. He and I discussed youth science education and how we each strove to involve youth of our respective countries in the space program. He was personally involved in youth education in Bulgaria.

The next day I had the pleasure of visiting a Soviet Air Force museum located at the air force base at Monino. The air base, which dates back to pre- World War II and was a part of the Gagarin Flight Academy, had been shut down. It featured an extensive story of Soviet aviation, beginning with the early 1900s, and continued through the most recent aircraft. The day I visited coincided with the visit of a member of the famous Sikorsky family. He, too, was an American. His father was responsible for influencing the design of many aircraft in the Soviet Air Force. He spoke fluent Russian and was warmly received. The tour covered six or seven hangars of prototype and first production aircraft, including the entire MiG jet family which became famous during the Korean War. Also included was an extensive collection of over 100 aircraft, including the world's largest helicopter, world's largest transport aircraft, the famous Soviet SST, their version of the XB-70, the Bear bomber and a DC-3 that was built in the Soviet Union during World War II under a licensing arrangement with Douglas Aircraft Company. I had the opportunity to fly for 30 minutes in the back seat of the MiG 29, their newest high performance jet and the prototype test aircraft of their first line fighter. It was comparable to a ride in an F-15. The only English spoken by the Russian pilot, Shasta, that I could understand was, *"This is —— hot."* I think he had been around too many U.S. naval aviators.

I visited the Young Pioneer Palace which is dedicated to the education of young people ages 10 through 13. It can best be described as a community education center. Enormous, it had a staff of 100 plus government employees and an annual budget of $5 million, U.S. . Eighteen hundred youngsters visited the palace two days a week for two-hour learning sessions. I was greeted by Boris Gregoryevidch, head of astronomy and astronautics. The Young Cosmonaut organization is part of his bailiwick. Many of his staffers had worked together since 1962. Specialties offered at the palace included airplane and rocket model construction, photography and radio station communications. They taught Morse Code and operated their own small planetarium. One room was dedicated to aircraft simulators, most of which were active. This was the department where the Young Cosmonauts made their home. Their regimen included physical training, a lecture on orbital mechanics by a professor from the University of Moscow, and the highlight of the visit, a simulated mission to Jupiter. The mission was conducted in the planetarium, modified to include the consoles and flight deck of a futuristic spacecraft, crew assignments and passenger section. It was done quite well, with interaction among the crew members and the consoles, and visuals interspersed. They had an excellent computer program with a timeline that triggered the various actions required of the crew. Upon completion, they celebrated a successful mission and exchanged gifts. I talked with many of the young people who in most cases spoke some English. I learned that two of the Young Cosmonauts who visited Huntsville a few years ago had gone through the same program and were graduates of that particular club.

Two of their graduates had applied for cosmonaut training. One was turned down and another was still being considered. Many of the graduates of the astronautics and astronomy department had gone to other specialties of work in the Soviet space program. Like Space Camp, they too were waiting for their first graduate to fly in space. We discussed potential cooperative efforts including communication satellite linkage between both facilities, a space station or Mars spacecraft design competition for the

Space Station or a Mars spacecraft and design competition for the next generation of space vehicles.

In 1990, the head master, cosmonaut Solvoyov, and some associates visited Huntsville's U.S. Space Camp and NASA. It was a great exchange of ideas. During the visit arrangements were made to conduct a joint project between the palace and Space Camp via satellite. It was exhilarating to listen to kids from two countries talking, some in English and some in Russian, about their mission to Mars.

In the '90s the U.S.-Russian space relationship was reunited and we began flying Russian cosmonauts on the shuttle and soon we were aboard their Mir space station. Today, we enjoy a strong relationship as American astronauts fly on a Russian rocket to rotate our astronaut crews to and from the International Space Station

A highlight of the Space Camp program in the 1990s was bringing foreign students to the camp and seeing the interaction between the young people from different cultures. Beginning with Russian children and teachers, we grew the program to include more countries and participants. International Space Camp hosted teachers and students from 30 countries who trained together and competed among themselves to form lasting relationships.

"Red Star in Orbit" the first USSR travelling space exhibit in the U.S. was negotiated by Buckbee and opened in Huntsville

In the late 1990s I began making trips to China. I had an opportunity to meet officials in the Chinese space program and was briefed on their future spaceflight program plans. Legend has it that toward the end of the 15th Century, a Chinese man named Wan Hu attached a bundle of rockets to a kite-like device, lashed himself firmly onto its seat and ordered the rocket fuses lit. There was ignition, a huge detonation and both Wan Hu and his craft took off. Legend does not tell where the spaceman landed. His supposed exploit has been a popular claim known by all of us in the manned spaceflight business that the first space man or "taikonaut" was from the land of China.

In October 2003, China officially did it. They became only the third country in the world to launch a human in space. Yang Liwei, a fighter pilot turned astronaut, became the first taikonaut to fly aboard the Shenzhous 5 (Shenzhous means "divine vessel"). Lt. Colonel Yang Liwei, at age 38, was a pilot with two decades of flying experience. He had flown for the People's Liberation Army Air Force since 1983. The 5 feet, 6 inch, 160 pound space flier came from a family of teachers in Liaoning province in China's industrial northeast. He joined the air force at the age of 18 and graduated from the No. 8 Aviation College. He entered the astronaut program in 1998. He is married to Zhang Yumei, who also served in China's space program. Yang was selected over two other air force pilots who had trained with him as astronauts. The Chinese Space Agency did not inform the public who would be the first astronaut from China into space, shades of how we announced our first Mercury astronaut, Alan Shepard. Another Chinese space flier—especially liked by the younger Chinese—was believed to be too Westernized and was not selected by the space agency. He wore American-made jeans, listened to rock music and loved Big Macs and Coca-Cola.

The Chinese press asked Yang if he saw the Great Wall from space as did U.S. astronaut Gene Cernan. When Yang answered, *"No,"* the Chinese press made some unkind and disparaging comments about his visual skills. Seeing things on Earth from space is not easy and is greatly affected by the orbit of the spacecraft and the shadows on the Earth-based object.

The Chinese will not admit they received help from Russia, however, it doesn't take a rocket

Artwork by Cosmonauts Alexei Lenov and Andrei Sokolov was a part of the "Red Star In Orbit" exhibition

scientist to determine their spacecraft and other technology have roots in Russia. Their spacecraft is a slightly modified, three-man Soyuz. It is launched on an up-rated *Long March* rocket from northwestern China. They recover their spacecraft by landing in Inner Mongolia with retro-motors firing from the spacecraft to lessen its impact before landing. Many Russian scientists and technicians found their way to China after the collapse of Mother Russia and the end of the Cold War. The Chinese taikonauts trained identically to the Russian cosmonauts and even went to Russia for some of the specialized training. According to Russian cosmonaut Valentina Tereshkova, the Chinese could not have accomplished the feat alone. Chinese taikonauts spent months undergoing rigorous physical training at Star City. There, they experienced the high temperatures and body-wrenching gravitational forces of takeoff and landing, and the weightlessness and discomfort of thickening blood in zero-gravity. Their plans for the future are similar to Russia's timeline for manned spaceflight in the '60s and '70s.

China began its quest for manned flight into space with research efforts dating back to 1958. They started a manned program in 1970, selecting astronaut candidates who were tested and trained but never flew. The program was abruptly terminated shortly thereafter.

In working with the China Society of Astronautics, I learned that Chinese children wanted to experience U.S. Space Camp. With the assistance of many people in the U.S. and China, several groups of Chinese children ages 9 to17 have attended Space Camp. The program continues to be a highly successful adventure for Chinese children.

This interest in space science education has stirred a movement to develop a Space Camp program in China. I am working with Chinese officials to open Space Camp China to motivate and excite children about space exploration. The initiative proposed will promote international cooperation in the peaceful exploration of space, similar to the initiatives that established Space Camps in Japan, Belgium, Canada and Turkey. I am hopeful such programs will encourage NASA officials to begin discussions with Chinese space officials regarding cooperative space projects.

Chinese Taikonaut Yang Liwei, China's first space flyer, with back-up crew members

Considering the museum setting that inspired Space Camp, it was natural we began talking about a place where the M7 astronauts, as well as those who flew in subsequent programs, could tell their stories. That place became known as the Astronaut Hall of Fame. A big supporter of the facility was Henri Landwirth, one of Shepard's closest friends. They had met when Shepard and the guys decided to move out of the government furnished housing (Hanger S of Cape Canaveral which they shared with a colony of apes). It was bad enough to follow apes into space, let alone share quarters with them. Certainly, it was a situation fertile for jokes. It was not uncommon to hear about NASA training this special breed of astronauts to feed chimps in space. Flight surgeon Dr. Bill Douglas added to the yarn by naming the astronaut's dining room, the Small Animal Feeding Facility.

Belgian-born Landwirth was the innkeeper of the Holiday Inn on Cocoa Beach. The guys referred him as "double-up Henri," implying he might sell a room to more than one person. Henri's reputation as a grand host and supporter of the space program was well known by all who visited the Cape—from contractors to the press corps. Alan and Henri began a relationship that continued for over 40 years. *"Henri became a close friend and in later years he became almost like a brother. He was and still is a great friend of all of us, particularly us Mercury guys. If you ask him, he'll probably say he raised us from wild fighter pilots to those crazy astronauts."* Henri was also the founder of "Give Kids the World" a place terminally ill kids would go and visit Mickey Mouse and Disney World. Henri donated thousands of dollars to support this very special place for terminally ill children. He enticed Holiday Inn, Disney, NASCAR and others to add big bucks to build a kingdom for kids who needed to have fun and be carefree, even if for a little while.

I knew of Henri from having stayed at his hotels during launch trips to the Cape. We became friends after Alan, Wally and I began to talk about a place to honor the M7. Once we hooked up, it was obvious we had the same feeling about the brotherhood. I remember talking to Henri and he told me, *"I know all these guys by their first names. I know their wives and families. They don't know how famous they are. We need to preserve them, honor them, and share them with the kids and brag on what they did. They should have a place of honor where everyone, today and in the future, can come and visit. Like Orville and Wilbur, these guys were the first!"*

As we continued our efforts to build the hall of fame, it became clear we should include a Space Camp as a part of the package. It just made sense, "Come see America's first space explorers and watch future astronauts train." After a few more meetings with Henri, Alan and Wally, it was clear a partnership would be formed between M7 and Space Camp. I received a letter from Henri in 1987 that stated in part, *"When the M7 Foundation was formed, our goals and objectives were the same as Space Camp. By joining forces, we can further the legacy for young Americans to explore their talents in science and technology. With the support of the pioneers of our space program, it is the most exciting venture of which I have been involved. My special thanks to Wally Schirra for his vision of bringing us together and Alan Shepard for devoting so much of his time and energy to making this partnership successful."*

I knew all the astronauts who flew in the Gemini and Apollo programs and had learned in order to get them involved in raising funds we needed to change the Astronaut Hall of Fame to be the story of all U.S. manned spaceflight, not just Mercury. I wrote a letter recommending to the M7 that we expand the Hall of Fame by honoring an additional 25 astronauts from the Gemini, Apollo, Skylab and Apollo–Soyuz programs. The name of the foundation was changed to the Astronaut Scholarship Foundation and for the first time, non-Mercury astronauts were named to the board. Today, the foundation has expanded the program even further by inducting space fliers from the Space Shuttle program.

Interior view of U.S. Astronaut Hall of Fame honoring Mercury 7

As we were opening the Astronaut Hall of Fame, a call went out to all the Mercury guys to send in their stuff and a new means of competition emerged. We received a shipment from Schirra and curiosity caught Shepard who said he'd like to see it. I took him to the temporary storage area where a sign on the door read, "U.S. Astronaut Hall of Fame, Lieutenant Commander Walter M. Schirra, Jr., Memorabilia Archives" with a photograph of Schirra. All the space artifacts we'd collected were contained in long rows of shelves—*some* of it belonged to Schirra. When Shepard saw the sign he said, *"What's this all about?"* He opened the door and looked in and I said, *"Welcome to Wally's collection."* You should have seen Shepard's face. His comment was *"Holy ——! Where in the ——did Wally get all this stuff?"* It was a gotcha Schirra enjoys to this day.

Shepard constantly referred to the fact that all the M7 guys were going to be asked to turn in their memorabilia, including things they'd brought back from space or the moon. He always joked that he'd been saving this jock strap he'd worn on the moon. It was a standing joke anytime we had a conversation regarding the Astronaut Hall of Fame. We held a roast for Shepard at Bernard's Surf, a restaurant on Cocoa Beach made famous during the M7 flights. This restaurant was patronized by all the astronauts, plus NASA followers. My roast was a gift; a special private supporter. I shopped around and found the smallest boy's size supporter I could find and I sprayed it with gold paint to the point it became stiff. I mounted it on a pedestal in a Tiffany-style, all-glass display container. On the jock strap was printed: "Rear Admiral Alan B. Shepard, Jr. USN, Naval aviator, test pilot, and Mercury 7 astronaut, Moonwalker, the Apollo 14 flight insignia with admiral stars and size XXL. The plaque read:

SHEPARD ARTIFACT

A device worn mostly by the male species to hold portions of the body in place while experiencing accelerating forces and sudden stops during spaceflight. NASA issue SN 001 worn by Alan B. Shepard. Jr.—the link between monkey and man – May 5, 1961, aboard MR-3, sub-orbital space flight, and January 31, 1971, aboard Apollo 14 lunar landing mission. Archivist note: Artifact

Space artifact, bronze replica of private supporter worn by America's first astronaut, Alan B. Shepard, Jr., presented at roast held at Coca Beach, FL

incorrectly labeled as XXL. Records show subject astronaut actually measured small or short.

But, of course, Shepard and Schirra never needed an audience to toast each other:

Shepard: *I think the way the spacecraft evolved was when Schirra showed up and he was lean, mean and skinny, relatively. We had the Mercury spacecraft and that was enough to get him off the ground. Then after a couple of years he gained a little weight and we had to go to the Gemini and the Titan. In the final analysis he had gained so much weight that we had to go to the big Saturn and Apollo vehicle to get him off the ground.*

Schirra: *During that time we developed Grecian Formula to cover Shepard's head! The real story is the Soviets launched Sputnik and Laika, the dog. We launched monkeys and chimps; the Soviets launched Yuri Gagarin for one orbit. We were devastated. Our next launch was to be a chimp. But we got thousands of telegrams from the American Society for Prevention of Cruelty to Animals (ASPCA), so we launched Shepard instead! And that's the REAL story.*

The last four remaining members of the M7 with the life size sculpture of Alan Shepard

Though it was 1993, Cocoa Beach residents thought they were back in the '60s when they saw a bunch of astronauts parading down the beach in Corvettes. Through Shepard's friend Jim Rathmann, Indianapolis 500 winner and Chevrolet dealer, 25 vintage Corvettes were provided for the space fliers to ride. Shepard was in the lead of the largest gathering of Mercury and Gemini astronauts in the last decade. Hundreds of people lined the parade route, cheering, waving flags and snapping pictures. Many who had watched these brave men lift off in those early rockets walked up to shake their hands and thank them for their service to America.

The parade kicked off a gathering of the surviving Mercury and Gemini astronauts at the Astronaut Hall of Fame. Allen Neuharth, founder of USA Today, involved his Freedom Forum as a sponsor of the astronaut event. The Freedom Forum is a nonpartisan foundation dedicated to free press, free speech and

Gemini astronauts inducted into Hall of Fame in 1993

a free spirit for all people. He hosted a bash for the entire crew at his residence, a renovated log cabin in Cocoa Beach called Pumpkin Center. He does his writing there in a beachside tree house overlooking the Kennedy Space Center launch pads. Neuharth has been somewhat of a space cadet, covering the astronauts in the early days of Cape Canaveral for his former paper TODAY, later named FLORIDA TODAY. That's when he and Shepard became good friends. They became such good friends, Shepard wound up taking a microfiche copy of the front page of Neuharth's paper to the moon on Apollo 14.

Neuharth would later commission a bronze heroic-size statue honoring Shepard as the first American in space and one the 12 humans who walked on the moon. The statue honors him not only for his space exploits but for his vision that led to the formation of the Astronaut Scholarship Foundation and the establishment of the U.S. Astronaut Hall of Fame. The statue is titled, "The Spirit of Space" and depicts Shepard wearing a Mercury space suit.

The Hall of Fame ceremony featured tributes by the M7 brotherhood, Carpenter, Cooper, Glenn and Schirra. At the press conference, Schirra commented, "That took a lot of brass." Glenn corrected him saying, "Wally it's made out of brass." Schirra replied, "John, you don't understand a joke when you hear one."

Shepard daughters Juliana Shepard Jenkins, Laura Shepard Churchley and Alice Shepard Wackermann-participated in the ceremony. Robert L. Rasmussen, director of the National Museum of Naval Aviation in Pensacola, Florida crafted the sculpture.

In honoring Shepard, who served as a Freedom Forum trustee from 1993 until his death in 1998, Neuharth said, "Alan Shepard's legacy as a pioneer in American spaceflight is secure. We hope that this memorial will also serve as a monument to his legacy as a patriot and his firm belief in the principles of freedom upon which this country was founded."

In January 1995, I sent a message to Alan and his wife Louise on the occasion of their 50th wedding anniversary. Here's an excerpt:

> Little did I know when I met Alan Shepard, there was a classy lady who is truly responsible for Shepard ever achieving anything. No doubt she is responsible for him being selected to become one of America's heroes in the now famous Mercury "Right Stuff" astronaut program.

Shepard struggled and through Louise's efforts was given the opportunity by his fellow astronauts to become the first Mercury astronaut to fly on a rocket. The Icy Commander accepted the challenge and proceeded to become the link between monkey and man, and completed the mission successfully, but not before wetting his pants and embarrassing Louise, the family and the entire nation. After several successful Apollo missions to the moon, Shepard decided it was safe for him to go, so he again used the influence of Louise and got assigned the command of the Apollo 14 mission, sometimes referred to as the Shepard golf outing. After returning from that historic mission and embarrassing every golfer on planet Earth, he looked around for a new adventure.

Shepard by now perceived himself as something of a national hero. He decided that he would develop his own farm team of special people who would follow in his footsteps and become the next generation of Alan B. Shepards. Further, he claims to have started SPACE

CAMP, which in his mind was designed to create little Alan Shepard's who would be taught to be cocky, arrogant, swaggering, happy-go-lucky prospects for the future astronaut corps. And lo and behold, 200,000 American youngsters graduated from the now famous Alan B. Shepard Farm Team for Astronauts.

Shepard had always wanted a book written about him and although he was unable to find anyone interested, Louise came to his aid and convinced the Turner organization to produce not only a book but also a television special. Moon Shot became a best seller and the television special was a hit. Shepard thought that he was the reason for the success of the book and television special. He assumed this because he believed that his quotes and on-camera presence were so powerful that he personally made both the book and the television production successful. The truth is the authors and producers had to fabricate subjects, accomplishments, and comments that would make Shepard look and sound good. My source for this information is Schirra. Schirra was quick to point out, Shepard really never had much going for him and each of us had to help keep him in the

Liftoff of Shepard and Company on Apollo 14

program. If it had not been for Louise's influence, Shepard probably would not have made it. In regard to the book and television production, according to Schirra, he was actually the one who ghosted most of the good material that was used and consequently, made the project quite successful. Schirra, who is not one to accept accolades, nor brag about his accomplishments, volunteered some other information that is probably not widely known. The publisher and producers had offered to feature Schirra as the leading character and he turned it down and suggested Shepard be given the attention.

Having said all of that, I wanted to be sure that the classy lady Louise got the credit she deserves.

What an exciting 50 years you two have had! I congratulate you both on your 50th anniversary and I wish you many, many happy and rewarding years. Good luck, good health, happy landings and keep your powder dry.

The Skylab astronauts who were inducted into the Astronaut Hall of Fame came to Huntsville to celebrate the 30th Anniversary of Skylab. Moderator Buckbee, on the left, introduced Joe Kerwin, Paul Weitz, Alan Bean, Owen Garriott, Jack Lousma, Gerry Carr, Ed Gibson and Bill Pogue. Absent was Pete Conrad the only deceased crew member.

Wernher von Braun spent 20 years in Huntsville, Alabama, following his dream of flying man into space to explore and discover. He built a team of rocketeers and left a rich legacy. But his legacy didn't end in Huntsville. In 1970, he moved to Washington, D. C. and became the founding father of the National Space Institute, a not-for-profit society promoting future space exploration in the public arena. The institute wasn't the chief reason for relocating, however; the main reason was to lead a future planning team at NASA headquarters responsible for developing a blueprint for the next 25 years of manned spaceflight. It was an exciting plan that included shuttle craft, space station, lunar bases and manned Mars expeditions. By 1972, von Braun had completed the plan and presented it to a seemingly indifferent audience. Congress and the White House were not interested. The passion for flying in space had waned as Vietnam and other problems had escalated.

Von Braun retired from NASA later that year and went to work for his long time friend Ed Uhl who ran Fairchild Industries in Germantown, Maryland. As vice president for engineering and development, von Braun had a serious interest in communications satellites—an area of expertise for the company. Von Braun was keenly interested in applying satellite technology on behalf of parts of the world where direct communications to the user would help educate the illiterate. Once again, von Braun proved the validity of space technology for an improved quality of life on Earth. His cross country travels and speaking engagements showed von Braun in his element, marketing and selling his new ideas as to how space technology benefited mankind.

Traveling became fun again for von Braun. Every day he practiced his favorite hobby, flying, having access to his own twin engine aircraft that he landed on an airstrip in the front lawn of his office building. When he was not flying on business, he made frequent weekend trips to Cumberland, Maryland to fly his beloved sailplane Libelle, having become a licensed glider pilot in Germany at the age of 20.

Von Braun didn't have much opportunity to enjoy his new line of work, his family and his accomplishments. After a prolonged illness, he died in Alexandria, Virginia on June 16, 1977. He was 65 years old.

Von Braun rocket team members gather at museum's Saturn V in 1987 in memory of von Braun who would have been 75 years-old

President Jimmy Carter paid tribute to von Braun stating, *"To millions of Americans, Wernher von Braun's name was inextricably linked to our exploration of space and to the creative application of technology. He was not only a skillful engineer, but also a man of bold vision; his inspirational leadership helped mobilize and maintain the effort we needed to reach the moon and beyond. Not just the people of our nation, but all the people of the world have profited from his*

The 36-story tall Saturn V, displayed at Space & Rocket Center, the icon of the Rocket City, Huntsville, Alabama

work. We will continue to profit from his example."

NASA historian Eugene Emme said it best, *"Cruel fate denied Wernher von Braun the chance to buy his ticket as a passenger bound for an excursion in space—his boyhood dream and manhood goal. Because of Wernher von Braun, however, almost everyone has been brought to the realization that we have been passengers on a spaceship all along—Spaceship Earth. Posterity will not forget him...."*

I've attended memorial services for three astronaut crews: Apollo 1, Challenger and Columbia, as well as separate services for other astronauts. With the Apollo 1 fire in the spacecraft—consuming the lives of Grissom, Chaffee and White—it marked the end of NASA's innocence. It was sad the first time and it became even sadder when Columbia went down. In 42 years, we lost 17 brave astronauts in human spaceflight: Three failures in 143 manned flights. This nation mourned and honored them with appropriate and deserving national memorials.

NASA has declared the last Thursday of every January as a Day of Remembrance. NASA's three darkest hours took place around this time of the year: Apollo 1 fire on the launch pad on Jan. 27, 1967; the Challenger launch accident on Jan.28, 1986; and the Columbia re-entry accident on Feb. 1, 2003.

The space program is the best of America—men and women pushing the state of the art—striving to accomplish high risk, technical achievements on a new frontier. We do it and we let the world watch. Sometimes we fail and cry. But most of the time we succeed and rejoice. Today, the American people are proud to say it is *our* Shuttle and the astronauts are *our* American heroes, the aces of spaceflight. They go where others dare not. We are proud that there are courageous, brave and adventurous astronauts.

One segment of the NASA family often forgotten is the men and women behind the Shuttle

The Wernher von Braun memorabilia room at the U.S. Space and Rocket Center at his death, June 16, 1977. In 1968, von Braun appointed Buckbee as the curator of the von Braun collection who established this special exhibit at the Center

crews; those left here on Earth who must make the critical, calculated and timely decisions about the space transportation system. Scientists, engineers, project managers, launch directors, flight directors and managers—the unknown, less remembered men and women—must commit to and certify that millions of parts in the machine are ready and safe to put in motion to launch their colleagues into space.

Since Alan Shepard's Mercury Redstone flight in 1961, the men and women involved in manned spaceflight (human spaceflight is the politically correct term) have assumed far greater responsibilities than most of us realize. Although it is the computer age, this special breed must be disciplined and trained to monitor sub-systems so as to anticipate the behavior of the total system. The final decision is made by a few who must rely on information provided by many. These are the engineers and technicians from across America who have accepted the challenge to work in a program that is highly technical, risky and unforgiving. They are the best in the field, ready, willing and able to accept the pressures and responsibilities of the positions. They have been trained to expect the unexpected, second guess the systems, understand and then execute, if necessary, the contingencies available to them. Though not adventurers or risk-takers, they have accepted an awesome responsibility. When the vehicle and crew are launched and return to Earth, their work is not done. They must analyze the flight data and prepare for the next launch.

Why do men and women choose this line of work—pressure-packed and laden with major responsibilities? I am not sure, but we are indebted to those who have accepted the challenge of flying humans in space. In my view, they are heroes too. President George W. Bush said it best after we lost Columbia, *"Space travel is not an option we choose. It's a desire written in the human heart."*

Chris McAuliffe
Teacher In Space

Space teacher Christa McAuliffe and all the space teacher finalists visited Space Camp

In 1989 the M7 astronauts were asked questions regarding the loss of Challenger and its crew.

Were the Apollo 1 and Challenger failures caused by the same thing?

Slayton: *Apollo I was first in a series and Challenger was the 25th. I think it was a totally different cause and effect factor. You were dealing with a different kind of complacency. People were beginning to think they were dealing with a commercial airliner. That's where we were in the shuttle, but it was a totally different thing. We had some known problems to the same effect that the solid rocket had been giving us a problem. People knew it and identified it, but they hadn't identified properly or fixed it quick enough. Again, it was one of those things that was going to get fixed sooner or later; it just didn't get fixed quite soon enough.*

Shepard: *I'm not sure I would take that approach Deke, if you'll pardon my saying so. Certainly, as far as the types of hardware failures were concerned, there was some divergence between Apollo I and Challenger. I*

would suggest that it was not a criminal decision made here, or a lack of attention. I think we have to assume that everybody was reasonably doing his or her job well. But yet the final analysis of why the decision was made to go ahead in Apollo and Challenger really rests on the judgment of people. If the judgment is lackadaisical or complacent, then sooner or later something is going to happen. Getting away from the hardware terms and dealing with people, I think that lesson can be learned in the years ahead, not only in the space program, but also in everyday life. If you get too complacent and get back on your heels, sooner or later you're going to make a bad decision.

Glenn: *When we started out in the very early 60s, if anybody had told us we'd go a quarter of a century on all the flights without losing a single person in space, I don't think one of us here would've bet a nickel on that. We used to talk about rather ghoulish conversations early in the program at Langley about how many would be lost. I think all of us thought there would be at least one or two out of our group in Mercury, as the odds, they were not dissimilar to what we had in testing aircraft, things like that. So to think that we had gone all that time without losing somebody in space and something that's as far out on the cutting edge of science as this was, I think was a remarkable tribute to NASA and all the people involved, the engineers' thoroughness in which they did their job. All of us are here today, living to attest to how good they really were.*

Slayton: *I think it's important to recognize that whatever happens, good or bad, it's people. The whole thing is people. You talk about hardware failures and you have to remind folks that no hardware ever failed, it's always the people that failed. It wasn't designed properly or manufactured properly, checked out properly or wasn't operated properly. It all ends up back to guys like us. I think that's where this fits into the flight crew part of it...Whatever hasn't been solved when you come off the pad is left for the crew to take care of. That's probably the most important part everybody in this room played. That was our career field, solving everybody else's problems once we got to the bottom line.*

Cooper: *That's somewhat a matter of perception. You have people talking about safe transportation and when you look at all the statistics involved, look at the fatality rate in automobiles. It isn't a safe transportation system. We don't have a safe transportation system per se in space. I think that anything that is research flight-related, should certainly be considered research and judged accordingly. As Deke mentioned, you have to consider that whoever goes on that flight is really expendable if necessary.*

Schirra: *I hate that word expendable. What we flew on is now called an expendable launch vehicle, which we were as well, in lieu of landing on the runway, I suppose...Teamwork is the essence of what we are. We really work together. We started out discussing how each of us individually had roles in the Mercury program, but we always continually went on saying how we came together. I remember those Friday afternoons when we would debrief about where we'd been all week and fill in each guy's background with what we'd learned individually about some particular exercise. We all were equally trained for a mission. When we first started, all seven of us thought we were going on that first flight. The result of it was on each flight one of us was somewhere in a responsible place taking care of that guy in orbit or on the sub-orbital flights. I think that represents the difference between our group and other test pilots of that timeframe that talked about how they knew all about our space program. They did not, because they didn't have that great opportunity to work together with this group of seven as I did.*

Shepard: *You know where I was during your flight. I was in a typhoon in Japan. I was on a 6-knot auxiliary tanker in the middle of a typhoon.*

Schirra: *Running out of booze, I might add.*

Shepard: *You still owe me one on that one.*

Carpenter: *As time goes on, I think more and more about the unique nature of this group and the unique camaraderie that has grown and survived over the years. That's a wonderful thing and I think we're all different, but we're all sort of a cut of the same cloth. Among those similarities is curiosity. It's a driver for all of us. We brought back some early new truths about the realm of spaceflight. Without that curiosity I don't think any of us would have wanted to do what we did. We were curious about whether we could build a machine that would do this. We were curious about whether or not the human organism could do it. The strength of our curiosity is what made the risk worth the benefit. You know the risk of spaceflight is there, but the benefit could be derived by all of us and made us want to do it. That's what we share as a group.*

Glenn: *I agree with Scott on that...This idea of curiosity is the basis of all human progress. The curiosity and how you do something better... We now have in our lifetime the ability to move out into space in a new laboratory. NASA figures that every dollar spent on space has come back and benefited our economy about 7 to 1. Even if they're off by 50 percent on that, 3½ to 1 is a pretty good investment at that. We now have developed the space transportation system. We're still working out the bugs obviously, or Challenger would never have happened. We now have the challenge of going up there and being able to really develop the new materials and pharmaceuticals and all the things we now can do in this new laboratory of space and weightlessness. The observations you can make of all sorts of things above the atmosphere, back toward Earth and on out into space. The curiosity, yes, as Scott said, but curiosity implementing bigger purposes. I think we're all just proud we were able to provide some of those stepping stones along the way.*

Shepard: *For those of us looking ahead, the things that are going to occur in space are exciting for all the reasons that it makes everybody's everyday life a little bit better. You know I think when you strip away the excitement, what we have left is a level of technology that's been created by the space program, by the money, the efforts, the application of attention, the energy, etc. We have seen this level of technology increase as the years have gone by. Technology improves everybody's everyday life. We learn more about the planet and what will happen to our planet in the future. It also filters into the engineering society and the scientific society of this country. In many ways we've lost leadership in the world because of our high standard of living, the high cost of wages, etc. Many areas that are labor intensive we don't compete well. Therefore, we have to do what we do best. What we're doing best is increasing this level of technology to keep us at the forefront and the cutting edge of what's new, not only in space, but also in medicine and other areas important to our everyday lives and that will keep us leaders in the world. In order to do that in technology areas, we have to continue to interest young people to get involved as we did. As the engineers who helped us, as the scientists who helped keep this technology going and to keep the U.S. technology leadership position with the rest of the world.*

Carpenter: *It is clear to all the people of this world ...that the United States has lost preeminence in space. I think that is very dangerous to this nation. I think preeminence in space is a condition of our national freedom. If we do not get cracking on a vigorous space program, we will live to regret it.*

If you could do it again, would you accept another mission to space? If yes, would you approach it differently?

Schirra: *I did it three times and I found no place else to go. I'd like to start worrying about Spaceship Earth for awhile and I hope sometime we see the project that has been labeled "Mission to Planet Earth" develop into something that will occur in the future. It is dependent upon a space station and many nations, not only the United States.*

Carpenter: *When we get down to the bottom line, we're eventually going to run out of space on Earth and destroy our atmosphere if we don't get in gear and start improving thing. I think it certainly behooves us as a people of the world to really look towards exploring, to find other planets we can populate. That's a long-range thing, but eventually we're going to have to look in that direction.*

Shepard: *You bet I would. It's been a marvelous experience and I'd do it again in a minute. Would I do it in the same way? Absolutely. I'd do it with the same guys, the same engineers and scientists and politicians who helped us so successfully over the years.*

Glenn: *I agree completely with "would you do it again?" Of course, but I'm disagreeing with Al regarding would I do it exactly the same...I think we'd all maybe make some changes in equipment we took and things like that. Would you want the same results? Yes. I think we all take great pride in having participated in something for a national purpose. I hope our national purpose just continues so this isn't something that was a flash in a pan in history. It's something I think that has to be moving us on out because we now have the capability to do some research and things of benefit for everyone here on Earth. We can't lay that down. That would be a big mistake.*

Cooper: *Well, I think from my point of view, the goal was to just be a part of the program. I felt very privileged to be in the program. I always felt I'd get whatever flight was meant for me to get. I think initially, of course, we didn't have the goal of going to the moon until into the program. Then, of course, the moon became a real goal. The goal also was to develop the hardware, equipment and know-how and find out if we could really get to the moon.*

Slayton: *If you're talking about the Mercury program and if you'd do it over again in the same way, my answer is not no, but hell no!*

Shepard: *Ha, ha. I knew that was coming. Would you learn to speak Russian again? You spoke such good Russian too, Deke!*

Slayton: *No comment.*

After Challenger, we had the Columbia accident. What do you think caused that accident?

Schirra: *Now that I look back on it, what contributed to all three accidents to some degree was "go fever." You got to go, you got to go, and you have to keep flying. We got into trouble with Apollo 1 that way and with Challenger. There was no reason to launch on that cold morning. With Columbia, they ignored things happening to other parts of the shuttle. When you look at all three accidents, there was a reason—there were several reasons—they should never have flown. All of those things were present in all three accidents and could have been prevented.*

I went to the Cape with Jim Lovell after the Challenger accident. We were dressed in clean room smocks and admitted in the holding area with all the recovered parts from Columbia spread out all over the floor. They wanted us to talk to the crews; to try and get them motivated again and to forget the accident and get on with the return to flight. I talked to them about having gone to safety school and test pilot training and learning not to look for the Easter egg without having gone through all the other perimeters so you don't want to make up your mind too soon as to what the cause is. As we were walking around, a guy came up to me with two eggs in a plastic bag. We all got a laugh out of that and eased the tension. They settled down and went back to work. I heard later that was the biggest moral boost they had down there in months. It's a sad commentary when pressure makes you look for the Easter egg when, in fact, you should take your time. Now, they are taking more time.

"The Wright Brothers may have taken our nation to the clouds, but Alan Shepard took us into space. He carried the dreams of a nation, and he made our spirits soar...He opened a new world to us and sparked a thousand dreams. History will never forget him and neither will we."

Vice-President Albert Gore

I visited Shepard at Pebble Beach in 1998 just before his death. He was sick but he put on a good show. We went to lunch at the Pebble Beach Golf Club and several people came by to say hello. It was obvious the citizens of the Pebble Beach community were proud he was in the neighborhood. After lunch, he suggested a little sightseeing. I knew he was a different Shepard when he said, *"You drive, Buckbee."*

As we were pulling out of the clubhouse, he suggested we drive along the coast road. Following his direction, we tooled along until he asked to take a photo of me standing next to the famous Cypress tree overlooking the Pacific Ocean. Back in the car and still following his lead, it wasn't long before Shepard pointed to the right and said, *"Turn here."* Puzzled, I asked, *"But isn't this the golf course?"* He responded, *"Have you ever played Pebble Beach or been on the course?"* I hadn't. *"Well, you're about to go on the course and more importantly I want to show you our house from the number 6 green."*

So, I drove down the golf cart path, passing golfers and guests in their golf carts. Talk about getting some stares—and these stares weren't the stares earned by the celebrity status Shepard enjoyed. I commented something like I didn't think we should be out there, with a car, not to mention an Avis rental car. I had a fleeting thought about fines for driving on Pebble Beach Golf Course and would Avis pay it and sue me. Shepard said, *"Just keep driving, Buckbee. Don't worry, they know me."* We stopped at the number 6 green and got out. The admiral pointed and said, *"See that house right up there? Now you can say you saw Shepard's house from Pebble Beach's number 6 green."* I took a picture.

We climbed into the Avis and slowly made our way back to the road. Along the way, we waved at golfers, greens-keepers and security people. No one stopped us, but there were some choice glares. *"Just keep driving,"* was Shepard's calm reaction. We returned to the house and sat on the deck surveying Pebble Beach, the bay and the green we had just left. It was beautiful with blue waves breaking at the rocks and shooting up white water. It was like a movie scene, almost too perfect to be real and unlike any I'd ever seen. *"Now,"* said Shepard, *"you can see why Louise and I have come to love this place."*

We talked about lots of things; trips together, some gotchas, good times, funny times and even embarrassing times. Space Camp came up and he thanked me for getting him involved, like I just happened to let him in on a good thing. *"I liked coming there and seeing those little guys and gals having a ball playing astronaut. It's the next generation of us Mercury guys at boot camp. If there had been a Space Camp back when I was growing up, I would have gone and I would have won the Right Stuff award."* That was Shepard.

We said our good-byes and hugged each other one last time. I drove away, seeing him standing proud and tall in the front of his house in my rearview mirror. That was the last time I saw Alan Shepard. I miss him.

When Shepard took his last flight, truly escaping the bonds of Earth, a memorial service was held at NASA's Johnson Space Center. On that first day of August the colors were presented by a Navy color guard and strains of *Oh Danny Boy* hung in the air. The auditorium was full of Shepard's friends, family, the brotherhood and other peers. Henri Landwirth and I sat together during the service. Several people spoke, including former flight director Chris Kraft. Excerpts from Kraft's remarks follow:

We are here this afternoon to remember Alan B. Shepard, Jr., as friend and a colleague, a man impossible to replace. In 1959 the country decided it needed a number of men to fly their spacecraft in Project Mercury. Some of us from the Space Task Group were looking for some volunteers who understood the rigors of test flight and were accustomed to putting their lives in jeopardy every time they flew. Alan Shepard volunteered to be one of the men we were looking for. He came into this new challenge because in his mind spaceflight was the next step in man's quest to fly higher and faster, and as a Navy test pilot, this was his business. He didn't come into the space program looking for glory. As the first manned flight approached, the competition for being first of the seven astronauts became very intense and in any group greater than one, it was difficult. Bob Gilruth met with the astronauts and asked them to write a brief essay about the astronaut who should be selected to fly first, excluding themselves. With the advice and consent of his staff, he picked Shepard to fly the first manned Redstone flight. These were heady times and Shepard, already charged, doubled his efforts in training for the 15-minute flight. Unfortunately, Shepard's flight had to be preceded by the flight of a chimpanzee. Although we successfully launched and recovered Ham, the hot engine of the Redstone caused an early engine shutdown which caused an escape rocket separation and subjected the animal to 17 g's and gave the faulty impression that the animal had failed to perform. The demand of more Redstone testing, and the requirement to test a large number of animals on the Johnsville centrifuge, postponed Shepard's flight. Shepard was extremely disappointed and agitated—as was the Space Task Group. The real disappointment came several weeks later when the Russians launched Yuri Gagarin and recovered him one orbit later... After test flights of Redstone were completed, the test program of the animals was cancelled, MR-3 (Mercury Redstone 3) *was given the green light. After several launch attempts were cancelled because of weather, on May 5, 1961, Shepard was again ready to fly. NASA had decided to allow the media to cover the launch, including TV to cover the launch in real time, putting tremendous pressure on all concerned. But this decision turned out to be far wiser than all of us could imagine. The launch was scheduled for 7 a.m., but rain and clouds covered the Cape. We decided to press on and allow Alan to enter the capsule and wait for the weather to clear. The picture I had on my console of Shepard exiting the trailer at the pad with searchlights brilliantly shining on him and his silver space suit is still ablaze in my mind.*

It was one of the most exciting moments of our lives. We all waited and finally, the weather cleared and we picked up the count. At T-minus two minutes, the magic point in the count in those days, Dan White began giving me Alan's vital sign readings and he reported his heart rate nearing liftoff reached 180. He was not the only one whose adrenaline was flowing. I looked down to make some notes and I realized the microphone on my headset was missing. I started to have it replaced and realized it was still there but I was shaking so hard it appeared as a blur in my eyes. We all remember Shepard's exclamations during launch and his staccato reporting and marvelous descriptions of the Earth and the heavens he viewed from his lofty position. He performed exactly as the finest test pilot was expected to do and his entry and recovery were equally spectacular. The Navy helicopter retrieved Shepard and his spacecraft and history was made.

The reaction of the country to this sub-orbital flight was incredible and Shepard had become the hero that he had not set out to be. The world praised the U.S. for its success, but particularly our willingness to share it with the rest of the world as it happened, (which) was in stark contrast to the Russians ...who did not report their missions until they were successfully completed. To celebrate the astronauts' accomplishments, President Kennedy invited Shepard and his colleagues to the White House to receive the country's highest honor. The impact that Shepard's flight made on the country and the world was not overlooked by the president and within weeks he had challenged the nation to landing man on the moon and returning them safely to Earth before this decade is out.

Alan Shepard had become the nation's first astronaut and inspired the country to set out on one of its greatest adventures. Alan's contribution to his country, his commitment to the space program and his willingness to go the last mile to show it, are what made him to so dear to all of us...

John Glenn was next to speak and here's what he said about his friend:

Alan Shepard was many things, a patriot, leader, a competitor—a fierce competitor—a hero, and to us here today, a close friend. When we seven were chosen as America's first astronauts, Al stood out. I first knew Al as test pilots when we were at the Naval Test Center, Pax River, Maryland. But it wasn't until the Mercury program that I saw, firsthand, the determination and toughness of Alan Shepard. Al named his spacecraft Freedom 7, based on his view of where this program stood, the openness of it around the world and what we stood for. In fact, as he sat in Freedom 7 through another delay in countdown before his historic Mercury mission, he was very anxious to get going. We heard an anxious Al say, "Why don't you fix your little problem and light this candle!"

America has lost one of its true adventurers, someone whose entire life was dedicated to America's questing spirit. Scott, Wally, Gordon and I have lost more than a friend, we've lost another brother...

In closing, Glenn repeated *High Flight*, the aviator's creed:

Oh, I have slipped the surly bonds of Earth
And danced the skies on laughter-silvered wings;
Sunward I've climbed and joined the tumbling mirth
Of sun-split clouds—and done a hundred things
You have not dreamed of—wheeled and soared and swung
High in the sunlit silence. Hov'ring there,
I've chased the shouting wind along, and flung
My eager craft through footless halls of air...
Up, up the long, delirious burning blue
I've topped the windswept heights with easy grace
Where never lark, or even eagle flew—
And, while with silent, lifting mind I've trod
The high untrespassed sanctity of space,
Put out my hand, and touched the face of God.

Al is right now, today, experiencing that ultimate, all-time high flight. But here, we all will miss him very much. Louise, the family, our thoughts and prayers are with you.

Scott Carpenter spoke next:

It's been nearly four decades since I first met Al Shepard. Shortly after that, I came to know him early in our training when I rode in his Corvette from Langley to Johnsville for some centrifuge training. By the time we got to Johnsville, I knew I was in the company of a dedicated naval officer, a unique and capable aviator, a born leader and a serious competitor in anything he wanted to do. The placard on the dashboard of his Corvette saying, "Specially made for Alan B. Shepard," only added to my awe and respect for this multi- talented man. I never asked and never found out how he came by that special compensation from General Motors, but I always wondered.

For a long time, both Al and I lived with some names we weren't particularly fond of. His was Bartlett and mine was Malcolm. We fell into the custom of addressing each other by those names. In keeping with that custom, I will now say, my hat's off to you, Bartlett. Thank you for all you have done for the Navy, for the space program and for your country. It was an honor and a privilege to have known you.

Following Carpenter was Wally Schirra:

… I would like to recommend that we remember him as Rear Admiral Alan Bartlett Shepard, U.S. Navy, Yankee from New Hampshire, Naval Academy graduate, black shoe— Cogswell in the Pacific, brown shoe—Naval aviator aircraft carriers, test pilot, astronaut and a very generous man… As we went through our training together, I saw Al Shepard stand out—I made the mistake of voting for him on that peer thing, now that I think about it (laughter).

I could see Al becoming a leader. We had the Mercury program going and the time came and Deke Slayton was to fly. We remembered him here, not too long ago. It's tough to come back again. Thinking of Deke, if he was going to be grounded, we better have him in charge. We didn't want an admiral—excuse me admiral (referring to an admiral in the audience)— *coming down to take care of us. We wanted one of our own. After Mercury, Al was due to fly in Gemini but was grounded with Meniere's Syndrome, grounded for almost a decade. Deke asked Al to help him out in the astronaut office. We had two great leaders. We've lost two great leaders. We're down to four now. We lost Gus on the launch pad. But that brotherhood we had will endure forever. An interesting part of Alan Shepard that came to light when he got the moon finally, and he hit that golf ball for miles and miles. My last recollection of Alan hitting a golf ball was on Kauai on the 6th green on the ocean's edge. He drove a ball off the edge into the Pacific Ocean and said, "That one went miles and miles and I didn't have to take a mulligan." We will remember Alan B. I was called Sky Ray by him… Bye-bye, Al.*

Gordon Cooper spoke next:

I sort of had a message for Al today. We were in the first group of six contestants; started out in Washington, spent a whole week of mental, written exams, went on to Lovelace Clinic. That was very interesting. They had been told to weed out as many of us as they could. We were the first guinea pigs to come through. They did weed out a lot of people and they were brutal. It was a real adventure. Then we went to Wright Pat, went through the physiology tests. They vibrated us, dunked us in ice water, sealed us in soundproof chambers and roasted us. We survived that. Then in April 1959, we were brought out on a stage like this and announced to the world who the first seven were. Boy, there were lots of cameras and lots of reporters. That was the first time for most of us to be exposed to public relations. But we survived that, Al. We got through it. That was interesting. And we survived the Navy UTE training, survived being roasted in the Nevada desert together, and jungles of Panama. We survived the research and running on the Johnsville centrifuge. It was brutal, but it proved how strong and capable the human body really is. We raced many miles in identical Corvettes (laughter). *I'm sorry, Al, I didn't tell you that I changed my ratio in my differential* (laughter). *You weren't any less of a driver; I just cheated a little on you.*

We spent many long days and nights getting MR-3 ready to go. I was blockhouse CapCom. We were together on the communications systems during the launch. I'll never forget there wasn't a dry eye anywhere in the blockhouse when we launched. I had Wernher von Braun and Kurt Debus at my right elbow and they were excited to speak to Al. And you were my CapCom on my Mercury flight. Well, we had lots of good flying, lots of good tests and lots of good times together. Now, you are up there in that big hangar in the sky, doing a lot of good flying. We miss you, Al. And we will be up there soon, and try some of that flying ourselves. My best to all of you and we'll miss you, Al.

The next speaker was Ed Mitchell, Apollo 14 lunar module pilot, the only living member of the Apollo 14 crew:

Good afternoon. It is my privilege but sad duty to represent the Apollo 14 crew on this sad occasion. I know Stuart would have wanted to be here but his wonderful family has come over to be with us for this occasion. And thinking of Stu and Alan, I'm reminded of an event as we came up to the lunar surface. We came in sight of the command module, and we could hear Stuart saying, "Hello down there, fearless one: What are you doing down there? You seem to have lost a little weight since the last time I saw you." I'm somewhat sure last week that Stu might have repeated almost those same words.

I am privileged that Alan chose Stuart and me to go with him to the moon. He was a wonderful teammate, colleague, leader and boss. I remember on the final day of our simulation of the lunar landing before the flight, when the team had thrown virtually every malfunction they could find at us and we successfully got down, I turned to him and said, "Congratulations, boss, I think we are ready to go." And he smiled in that enviable way and said, "Yeah Ed, I think we are."

I want to pay tribute to him and what a privilege it was to serve with Alan Shepard. I'm sure that where it is that astronauts and explorers go when we depart this realm, Alan and Stu, Deke and Gus, Roger, Ed, Elliott, Charlie, C.C., Jack, Jim and Ron are pioneering and exploring and doing things that one day we will all be together again. Farewell, and thanks a lot.

Jim Lovell, commander of Apollo13, spoke next:

I first met Alan Shepard when he and I were part of the original Mercury candidates group for the Mercury program. Al offered me a ride in his Corvette from the hotel to the Dolly Madison House where we were briefed by Project Mercury. Of course, I knew of him because of his work at the Naval Test Center. I thought this is one naval officer who's got it made. I asked him how much one of these cost and Alan replied, "If you have to ask, you can't afford to own one."

Al was a natural-born leader, the type people want to emulate and follow. And sure enough, it wasn't long before a lot of us were driving Corvettes. Al's legacy as a pioneer in the space program is firmly etched in the history books. He was fiercely competitive and won the competition to be the first man in space. He had the motivation and perseverance to hang in there after being grounded for eight years; he finally flew as the commander of Apollo 14 and landed on the moon at a place called Fra Mauro. When that opportunity was denied me, I cannot think of a more appropriate person to take my place than Alan B. Shepard. Al's legacy in space is a thing of the past, a legacy future generations can reflect on when they want to reminisce about the early space heroes of space exploration.

But Alan Shepard left another legacy, one that didn't die with him, a legacy that is alive, growing and prospering. Al was the driving force behind M7 astronauts establishing the Astronaut Scholarship Foundation... Thank you, Al. We will miss you.

Not many weeks went by when I received word that Louise Shepard had passed away on August 25, 1998, as she was returning to the Shepard home at Pebble Beach. She died just weeks after Alan's passing. One close friend observed that Alan had called her home.

The family held a simple and moving memorial service for both Alan and Louise at the Pebble Tennis Club on November 18, 1998. Several friends of the family spoke in the private ceremony held in the beautiful Pebble Beach setting.

Dorel Abbott, one of Louise's closest friends since middle school, spoke of their enduring relationship. *"Louise grew in stature over the years from a somewhat shy, introverted young woman to one who hobnobbed with celebrities, often staying at the Sinatra home, entertained at the White House and experienced great travel."* Abbott remarked how Louise watched her husband receive many accolades, all the while offering her support.

"I never saw her lose her temper, always gracious and humble, a perfect lady. We never had a fight, never a cross word. She was a devoted wife, mother and grandmother. She was the closest thing I ever had to a sister. I love her very much and shall miss her dearly," remarked Abbott.

Gene Cernan, commander of Apollo 17 and backup to Shepard on his Apollo 14 flight, kept in character and represented the brotherhood as he spoke of Alan and Louise Shepard:

"How do you talk about Alan and Louise Shepard? I've been searching for the words for a long time that adequately describe what they meant to us all. Alan had gone where I never dreamed of going. I remember standing next to Alan Shepard, 10 years later, after the same crazy guy that rode that Redstone—but this time standing next to that monstrous rocket ship, Saturn V that was going to take him to the moon the next morning. I don't know how he felt, but I felt extremely humble, standing next to Alan Shepard who was going to ride that Saturn V and Apollo 14 spacecraft to the moon.

"In the course of that time, I got to know your mother, Louise. How do you find words to describe the lady who became the first lady of the space program? I can't. But that's true. She became that and remains so today.

"Someone who was very close to me once said, "If you think going to the moon is hard, you ought to try staying home." Some of you ladies out there know exactly what I'm talking about; how hard it is to stay home. We haven't thought about that, but I guess it is pretty hard to stay home. Louise Shepard had to stay home while the world marveled and revered Alan Shepard's accomplishments in space. But that same world was cheated because they never got to know Louise Shepard as many of us in this room knew her. Because to know Louise Shepard was to admire her, be her dear friend, respect her, to love her. But Alan knew Louise Shepard. He loved Louise. He didn't necessarily flaunt it, but Louise was Alan's best friend. Alan admired and respected her and most of all, Alan loved Louise Shepard.

"When Alan found himself in trouble, when he had medical problems and ran into all kinds of things no human being should have to endure, he turned to Louise. When Alan was looking for a hero, Louise would always be there.

"When Alan was challenged to reach further into the sky and needed more wind beneath

his wings, he turned to Louise. Louise was always there. And when those of us in the space program, in times of triumph and tragedy, needed to turn to someone, she was that champion. She was always there in style, charm, beauty and class. She played a major role, sometimes unheralded, in America's space program and the lives of so many of us.

"A few months ago, Alan took his final—certainly his highest—flight of his career. But this time, your mother didn't stay home. She took that trip with him and shortly thereafter she too, had the opportunity to reach out and touch the face of God, like Alan had done on so many occasions.

"Alan and Louise, you both will be sorely missed, but I promise you, not soon forgotten. The world is a better place for you having been here. And you taught those of us who follow in your footsteps the meaning of courage and commitment.

"I like to wish you, in the solemn Navy tradition, as you share the heavens together for eternity, that you always find fair winds and calm seas. Good-bye and God bless you all."

As the speakers finished, we were led out to the 17th green to witness the final flight for Alan and Louise. The Navy had a tradition of burying their dead at sea, always a distance from shore and out of sight of people. You might have guessed Admiral Shepard lobbied his Navy buddies to change this so his and Louise's ashes could be scattered in the cove within view of their beloved Pebble Beach home. Even after death, the Admiral was letting us know he was still in charge.

Memorial service for Shepards at Pebble Beach, 1998

As family and friends gathered, we heard jets before we saw them approaching from the south. There were four Navy F-18s flying slowly, low in the sky. As the formation approached, one pulled up and out of formation: the missing man formation, honoring a fallen naval aviator and his lady.

As the Navy jets faded away, two Navy helicopters came over the Shepard home and proceeded to make a wide circle turning south over the cove before assuming a hover position. These two giant Sea Hawk helicopters held their positions for a few moments. From one was released the ashes of Louise Shepard; from the other, the ashes of Alan Shepard. The Sea Hawks applied power and flew off into the distance. Alan Shepard expertly deployed Navy aviation one last time, orchestrating the most impressive Navy burial ceremony ever witnessed. He had to be happy with the outcome because it was typical Shepard—moving, memorable, unparalleled. When the ceremony concluded, I turned to my wife Gayle and said, *"That was classic Shepard."* Family and other close friends were also looking at each other saying, *"Yes, that was Shepard."* The Admiral left us something to remember. He would have been proud of how it impacted all of us present that morning. And as Deke would have said, *"Okay, José and Louise, you are on your way."*

First International Space Camp attended by teachers and students from twenty-six countries in 1990

Space Camp—Trainees assembling large space structure

The most valuable gift we can give a child is a careful education. This is where society has a prime obligation. To waste the time that a child has to acquire knowledge is an outright sin. **Wernher von Braun**

Shepard had the audience in the palm of his hand. There were about 500 kids present, ages 10 to 17, who had just graduated from the week-long U.S. Space Camp in Huntsville, Alabama. Shepard was describing his early days in aviation, how he became interested in spaceflight and then eventually became a Mercury astronaut. He told them Charles Lindbergh had been his hero when he was their age. As he finished, he asked if anyone would like to ask a question. One little guy's hand shot up and he asked, *"Admiral Shepard, why were you selected to be the first American astronaut to ride the Redstone rocket?"* Shepard's reply was, *"Because I was the best."* That was Shepard.

I was often asked why I started Space Camp. *"Well,"* I would answer, *"when I was a young guy, my sandbox was too small, my tricycle wouldn't go fast enough and my bicycle wouldn't fly. I felt I had an uneventful and dull childhood, so I started Space Camp. Today my sandbox is next door to the biggest collection of the rockets and spaceships in the world, the U.S. Space & Rocket Center."*

Space Camp—Multi Axis Trainer

I remember the summer of 1990 when Space Camp was bursting at the seams with young people attending from 50 states and 26 foreign countries. About 20,000 campers graduated that year. It was an exciting time to be around this energized generation that had set its goal of some day flying to Mars. Space Camp was another one of those great Wernher von Braun ideas. It was my good fortune to nurture and implement the concept. After von Braun left Huntsville in 1970, he often returned to visit the Space & Rocket Center. With me as tour guide, he would ask questions and test some of his marketing ideas as we walked through the museum and rocket park where the machines he created were displayed. During one of those visits, von Braun noticed a group of students diligently taking notes as their teacher rattled off key facts about the Saturn V moon rocket. After some discussion about student field trips, he looked in the direction of the students walking down the length of the Saturn V, hesitated a moment, and said, *"We ought to have some kind of program that is a continuing educational experience, like a camp. We have band camps, cheerleading camps, football camps and scout camps. Why don't we have a science camp?"* That day the vision was shared and I accepted the task of creating U.S. Space Camp and Aviation Challenge, two unique educational programs that have graduated over 500,000 young people. Von Braun's dreams are alive today in the minds of thousands of young people who have unbridled enthusiasm for science, exploration and discovery.

One of the main goals of Space Camp has been to motivate youngsters to math and science

Space Camp—Five-Degree of Freedom Trainer

Space Camp—Moonwalk Trainer

excellence, using the space program as the inspirational answer to, "Why?" Space Camp teaches leadership, teamwork, decision-making skills and self-confidence. The program challenges young people to turn dreams into reality. It is an intense experience.

The youth of America are our most valued resource. They are looking for their niche, and they are asking for our help. We must nurture and cultivate skills and imagination. We must challenge them to aim high and accept nothing but the best. Space Campers have made some memorable comments over the years. When asked by a team leader what he thought of Space Camp, one trainee answered, *"It was the first place I've ever been where it's cool to be smart."* Another, during a phone call home, was overheard saying, *"Hey Mom, guess what? All the nerds are here—I love it!"* During an interview, a reporter asked a camper what he would do if he ever became the head of NASA. The kid thought a minute and said, *"Well, I'd have a crew on board the space station. I would have a colony on the moon and I would have an exploration team going to Mars."* The reporter commented, *"Wow, that's bold thinking and expensive. I don't believe Congress would approve that."* The camper's response was, *"Your Congress might not, but mine will."*

But not everyone who comes to Space Camp wants to be an engineer or astronaut. I asked one young lady what she wanted to do and she informed me she planned to be a surgeon. When asked why she came to Space Camp, she replied, *"Because I may have to operate in space someday."* Another guy said he planned to be a professional golfer and make a lot of money and he thought being an astronaut might make a good second vocation. Shepard liked that comment.

It has been my hope to get young people excited about discovery and exploration. They should enjoy the same purposeful and energetic spirit that permeated this nation when Shepard and his buddies pioneered the space frontier and landed on the moon. The

Space Camp—Space Station Mobility Trainer

Space Camp—Manned Maneuvering Unit

American public likes being number one in the world and some of our greatest accomplishments can be traced to the Space Race. Throughout the world, the imagery of the American West extends beyond the boundaries of the Great Plains to the wide open spaces inhabited by our vigilant, rough and ready, space cowboys.

I remember a typical week at Space Camp for the little guys and girls (trainees) involved the launch of "cricketnauts" (often named Neil or Sally) in rockets each trainee had constructed. A tour through the center's Rocket Park, led by rocket engineer Konrad Dannenberg, an original member of the von Braun team, was like Rocket Science 101. Trainees had an opportunity to ask questions about every manned rocket the U.S. had ever launched into space, including monkeynaut Baker's Jupiter.

The big event was the mission, which took trainees on a simulated Space Shuttle flight. Each trainee performed as crew members to complete the mission successfully. This was the most dramatic and at times, the most nerve-wracking event of the entire week; the one event they would always remember.

Parents were into Space Camp as much as the children. It was a significant investment, to send a child to camp, airfare included. I remember the parent calling me mid-week, asking what was happening to her 12 year-old son. He had informed her he wouldn't be home Friday night when camp was over because he had been selected as payload specialist on the next shuttle mission. It just so happened that a real flight was scheduled to liftoff on Saturday of that week and the young man mistakenly thought he was going on the real thing. I thanked the mother for her call and informed her that Space Camp was very realistic but we were not flying campers on the real shuttle, yet.

How could I not remember the House sisters from Indiana? Carmella and Analda together attended a total of 17 sessions, plus their mother, Mariette, attended the adult program!

I watched Space Academy trainees as

Space Camp—Underwater Training

Space Camp—Mission Control

they practiced their extravehicular activity (EVA) drills, assembled a space structure, powered up SpaceLab experiments and learned how to fly the Space Shuttle orbiter to a safe landing. Catching up with the advanced Space Academy trainees was not easy. These guys and gals were everywhere. They were on campus for eight days participating in a very intense program that emphasized the International Space Station (ISS). This group of 20 team members trained hard all week for what was a very realistic 24-hour mission involving living, sleeping and working aboard the ISS. These high school aged students were treated to a session with Georg von Tiesenhausen, another member of the von Braun team who gave exciting glimpses of future space travel.

Three-time Space Shuttle astronaut Mike Mullane was the resident astronaut for one summer. Yes, there was a real astronaut living, working and teaching at Space Camp that summer. Mullane made it all believable for thousands of youngsters that summer as he shared why he became an astronaut, why they should study science and math, what's astronaut training, and most of all, what it's like to float in space, free of gravity, and ride that Space Shuttle three times.

Under the leadership of Deborah Barnhart, the commandant of Space Camp, the program was elevated to another level, adding exciting and realistic space and aviation training experiences. Team leaders Karen Perry, Tom Siranno, Taylor Jernigan and Paul Shanley made their mark as exceptional role models for trainees.

When questionnaires were sent out to prospective campers, they were asked their first and second choices of a profession. Popular selections were, not surprisingly, astronaut and fighter pilot. Aviation Challenge was a follow-on program conceived and built at a lake setting a mile away from the Space Camp training center. It became the home of the future fighter pilots, a "Top Gun wanna-be" campus. This was one of the new programs offered to youngsters interested in military aviation. It started with a strong male enrollment but soon was infiltrated by the opposite sex who filled about 40 percent of the trainee slots.

At the lake it was a kick to watch the first trainees go off the 100-foot tower on the "zip" line, simulating a downed pilot parachuting into water. It was all part of the survival training that included an all-night survival drill that not only challenged the trainees, but the staff as well.

I enjoyed checking out the trainees participating in the You Can Fly program. As I arrived at Signature Aviation, a four-passenger Aerospatiale Tampico piloted by an instructor from the University of North Dakota was just taxiing on the tarmac. Space Camp trainees climbed out and gave high fives to the instructor, ending their first flight aboard the single engine plane. They each had navigated, communicated and taken the controls of the aircraft. It was the encouragement many trainees needed to become licensed pilots or go into military flight training.

That particular week, a young fellow by the name of Nick Morgret from my hometown, Romney, West Virginia, attended Space Camp. While there, he had the opportunity to meet Alan Shepard. From his Space Camp experience he became interested in aviation. He obtained his private pilot's license while in high school and decided to pursue the study of aviation. I wrote several letters of recommendation to universities for him. He was accepted at all five. He chose Parks College of St. Louis University where he received his bachelor's degree in aviation science. Today, Nick flies for a commercial airline.

There were other exciting times—visits by President George Bush, Vice President Dan Quayle and U.S. Senators Al Gore and John Kerry—and the movie production *Space Camp* filmed in 1985. The latter brought actresses Lea Thompson, Kate Capshaw and Kelly Preston, and actor Tom Skerritt to the training center, along with producer Patrick Bailey and director Leonard Goldberg. The motion picture literally put Space Camp on the map in this country, and introduced the concept abroad.

Other celebrity visitors included Arnold Schwarzenegger, Willard Scott, Karena Gore, Chelsea Clinton, Hugh Downes, Nichelle Nichols, Ron Howard, Tom Hanks, Bill Paxton, and entertainers Bruce Springsteen and the group Alabama. All of the Mercury, Gemini and Apollo astronauts and every astronaut who walked on the moon—all twelve of them—paid a visit to the campus. Christa McAuliffe, the educator who died when Space Shuttle Challenger was destroyed, and her backup,

Barbara Morgan, visited with all of the teacher-in-space finalists.

U.S. Space Camp and the M7 had a very special relationship that joined "the first and the future." The relationship sprang from my friendship with the nation's original astronauts and it proved mutually beneficial to our missions. *"The way Alan and I saw it, it was a natural that we as astronauts wanted young people to follow us. I like to see success groomed to come along behind us. Space Camp is the best way I know of to entice children to be interested in what we did in the past,"* explained Wally Schirra.

The Mercury 7 Foundation was established in 1984. The original Mercury astronauts felt that somehow there should be a way to encourage youngsters to get involved in scientific or engineering pursuits. Eventually, they created a college scholarship program for deserving students who aspired to work in technology fields. *"We're not limiting technology to only space,"* emphasized Shepard. He further explained that by working with Space Camp, which addressed youth from fourth grade through high school, they increased their pool of motivated scholarship applicants.

Some of those scholarship recipients joined the ranks of the thousands of counselors and team leaders who were integral to the success of Space Camp; who gave up their summers to be the brother, sister, mother, father and best buddies to the campers. They were the glue that held the program together and made it so meaningful to the little guys and gals. Many of them have told me how fortunate they were to have taken part in the Space Camp experience. For many of those young adults, Space Camp prepared them to become true stewards of the American dream.

Working with young people was a responsibility I took seriously. Understanding that we have few chances to truly impact youngsters, I frequently offered comments during Space Camp graduation ceremonies, such as those that follow:

I've always dreamed of what it would be like to climb aboard a space vehicle and fly to another planet, land and become the first Earthling to inhabit cosmic territory.

I won't have the opportunity, but those of you in the "brotherhood of space campers," will not only have the opportunity, but I sense you will have the drive and determination to make it happen for yourself, your country and indeed, for all mankind.

Some in my generation will find those thoughts amusing, and downright fantasy, but I know you believe it can happen and that's what sets you apart from others. Space Campers are a breed of young people who have that special gift of an "I can do it" attitude. It's rare, but you can sense it if you have been observing young people as long as I have. You are the first, true, space flying generation. It is your lot to reach beyond your grasp, to think and prepare for the future and do it on a global basis with all peoples of the world, striving to explore space for peaceful purposes.

I can see it in your eyes and hear it in your voices, the excitement and anticipation of finding that secret at Space Camp that will enable you to acquire the edge, the advantage that will make you the best at what you want to do. You don't speak of it, you don't boast about it to your friends, but those of you who have lived the Space Camp experience know what I'm talking about. I'm counting on each of you to keep the dream of spaceflight alive. And when one of the brotherhood takes the first step on Mars, I'll bet we'll hear, "Hey guys, this is just like Space Camp!"

Dottie Metcalf-Lindenburger is one trainee who is keeping the dream alive. She is the first Space Camp graduate chosen for astronaut training. The youngest member of the NASA Astronaut Candidate Class of 2004, Dottie attended Space Camp as a high school freshman in the spring of 1989. Her interest in science and space exploration grew and she earned a bachelor's degree in geology from

Whitman College in Washington and was teaching high school science and astronomy in Vancouver, Washington when she learned about NASA's interest in professional educators as astronauts.

"The educator astronaut position had just been posted," said Metcalf-Lindenburger, referring to the NASA Web site. *"I got so excited. It seemed so perfect."* She pursued the opportunity, received the call and began astronaut training in June 2004.

NASA Administrator Sean O'Keefe announced, *"Dottie is just one of thousands of young people who has benefited from the tremendous learning experiences that Space Camp provides."* During remarks before the Alabama State Legislature, the same body Wernher von Braun challenged 40 years earlier to build the world's largest space museum, O'Keefe continued, *"I suspect in the future, more and more of our astronauts will have this Space Camp pedigree."*

I truly believe Space Camp became Dottie's stepping stone to the astronaut corps. And as an educator, she aims to use the excitement of space to ignite a passion for math and science in the classroom. *"A lot of kids aren't necessarily interested in science and math. But they do get excited about things like the Mars rovers—Spirit and Opportunity. I want to continue to build more connections with the education community to get them jazzed about studying science,"* she asserted.

Mercury Redstone model rocket launch at Space Camp Florida conducted by, left, Shepard; Chuck Hollinshead, Public Affairs officer, NASA Kennedy Space Center, FL; Henri Landwirth and Buckbee

"Through these doors enter America's future astronauts, scientists and engineers," states a sign I had installed over the door of the Space Camp Training Center when I was director. I knew one day we would have a Space Camp graduate become a member of the brotherhood. But, I had no idea it would be a female and an educator-astronaut. What a great accomplishment for Dottie Metcalf-Lindenburger, and an honor for Space Camp and all of us who have been associated with the program for over 25 years.

Another accomplishment hosted at the Space & Rocket Center was the 20th anniversary celebration of America's first moon landing held in July 1989 at the Space & Rocket Center, Shepard spoke before an audience of over a thousand Huntsville aerospace workers and Space Campers. *"I understand you had some chaps named Armstrong, Collins and Aldrin earlier and I understand you had a chap name Alan Bean here. Obliviously they saved the best until last. I would like to point out something especially significant to me about the Redstone and the Saturn V that you see here. Perhaps not all of you realize I was the only astronaut lucky enough to fly both the Redstone and Saturn V. They were both designed, developed, built and tested by folks here in the audience tonight and I want to flight. Three weeks later President John F. Kennedy made the decision that we are going to go to the moon and just eight years later man landed on the moon. Incredible. It is the story of technology, of dedication of people and it was all made in the good old U-S.-of -A. Because I am the oldest of the 12 astronauts who walked on the moon—I suppose I can take a little liberty to talk about that experience for a brief moment. The one thing that struck me the most, about being on the surface of the moon—a quarter of a million miles away from home—was our planet, looking back at Earth and seeing the beauty of Earth, the beauty of our blue planet. I was actually overwhelmed; tears actually came to my eyes, totaling surprising me, without any forewarning. To me it was an expression of something, which was obviously exciting for our generation, exciting for many generations. As a matter of fact, the moon landings were probably the most exciting things to happen in this century. We can all be proud to have been a part of it. We can think of those exciting years of the great Apollo program and being on the lunar surface and we can look ahead. Many of you here at Space Camp will be the space explorers of tomorrow. Where shall we go in the years ahead? Well, I'm not going to predict tonight whether it should be a permanent space station, a man landing on Mars or lunar colonization or perhaps all of these in due time. I think the important thing to remember tonight is we are celebrating this tremendous technology that we have in this country today, and knowing that technology must continue.*

Buckbee and Vice-President Al Gore at Space Camp

"We must continue to believe in technology because it helped us become the leading nation in the world and technology is going to continue to help us in our everyday lives," he shared. He repeated similar words years later in a series of interviews when he noted that space technology is important to society because it improves communications, offers new techniques, new materials and new ways of doing things. *"It's important we keep this tremendous momentum going that we achieved with space technology so that it makes the world a better place to live,"* he added.

As director emeritus and founder of U.S. Space Camp, I am inspired by the next generation of space fliers—the future pioneers who will keep the momentum going. I have seen them at Space Camp. These young people come from around the globe to explore their potential as the astronauts, scientists, engineers and educators of tomorrow. Who will be the first to set foot on Mars? Just ask these young people and they'll tell you, *"Me!"* or *"We will!"*

Moonwalker Jack Schmitt shares some humorous stories with Buckbee and Ernst Stuhlinger, chief scientist for the von Braun team, at Space Camp

Yes, dreams, imagination and enthusiasm are vital characteristics of pioneers worthy to follow in the footsteps of innovators such as rocket scientist Wernher von Braun and the Mercury, Gemini and Apollo astronauts. Lunar settlements, space colonization, missions to Mars, and many, many more projects await youthful vision, dreams and commitment. Keep the dream alive.

Whether we go to Mars tomorrow, or 25 years from now, the important thing is we have to continue developing space technology. We must continue space exploration. It is the legacy us old guys leave to you young guys. Take it, wear it well and do well.
Alan Shepard

Cast members from "SpaceCamp" the movie, pose with Lee Sentell, who as director of marketing for Space Camp set up a live broadcast from Space Camp to Good Morning America. Those on the show were, from left, Sentell, actors Tate Donovan and Kate Capshaw, Center Director Buckbee and actors Leaf Phoenix and Larry B. Scott. Currently, Sentell is the director of tourism for the state of Alabama.

Buckbee and Space Camp producer Patrick Bailey stand in front of the bi-plane that Bailey flew during the filming of Space Camp shot at U.S. Army Redstone Arsenal airfield

Vice President Dan Quayle and Guy Hunt, Governor of Alabama, observe Space Campers during visit in 1992

Space Camp founder, Buckbee presents graduating camper with certificate

Index

Exclusive double-sided DVD-Video^

The enclosed disc includes the following movies:

* 1972 Interview with Wernher von Braun
* 30th Anniversary Skylab documentary
* Alan Shepard "Gotcha" film
* "Reach for the Stars" Space Camp Documentary
* Apollo Panel discussion with Neil Armstrong, Buzz Aldrin, Alan Shepard, Eugene Cernan and Jim Lovell
* 2002 Mercury Astronaut Panel discussion
* "The Lighthouse that Never Fails" "Gotcha" film
* "From the Valley to the Moon" documentary
* "The Flight of Freedom 7" documentary
* "The First" documentary

* Apollo 12 "Gotcha" film

^ DVD is NTSC Region 0

About the Authors

Ed Buckbee, an author, lecturer, space expert and director emeritus, has been associated with the U.S. space program for four decades. Buckbee began his NASA career in 1959 when America's first Mercury astronauts were selected. He attended the launches of Alan Shepard and John Glenn and was present when the Apollo astronauts lifted-off for the moon landings.

In 1961 he transferred to the newly formed NASA's Marshall Space Flight Center where he worked for rocket scientist Wernher von Braun. As a NASA public affairs officer, he met and worked with all the astronauts who flew the early Mercury and Gemini space missions and Apollo moon walkers.

In 1970, he was selected by Von Braun to be the first director of the U.S. Space & Rocket Center. Buckbee assembled and managed the world's largest space and rocket exhibition and founded the highly successful U.S. Space Camp and Aviation Challenge programs. Over 500,000 students and teachers from seventy countries have been inspired and motivated by attending programs Buckbee developed. He teamed up with Wally Schirra and Alan Shepard in establishing the U.S. Astronaut Hall of Fame and developing Space Camp programs world-wide. www.air-space.com www.therealspacecowboys.com

Wally Schirra, Jr. was one of the original seven astronauts chosen for Project Mercury, America's first effort to put men in space. He was the only man to fly in America's first three space programs: Mercury, Gemini and Apollo and has logged a total of 295 hours and 15 minutes in space. They were the Mercury 7. Wally Schirra passed away in 2007. www.wallyschirra.com